it

The Architecture of Existence

it

The Architecture of Existence

Gary

Volume One

JANUS PUBLISHING COMPANY
London, England

First published in Great Britain 2008
by Janus Publishing Company Ltd,
105-107 Gloucester Place,
London W1U 6BY

www.januspublishing.co.uk

British Library Cataloguing-in-Publication Data
A catalogue record for this book is available from the British Library

Paperback ISBN Vol. 1. 978-1-85756-648-2
Paperback ISBN Vol. 2. 978-1-85756-653-6
Paperback ISBN Vol. 3. 978-1-85756-658-1
Paperback ISBN Vol. 4. 978-1-85756-633-8
Hardback ISBN   978-1-85756-668-0

Cover Design: Gary Hansen

Printed and bound in Great Britain

To whom **it** may concern

"if you can get the right book at the right time you taste joys not only bodily,
physical, but spiritual also, which pass one out above and beyond one's ...
self, as it were through huge air, following the light of another man's
thought. And you can never be quite the old self again. You have forgotten
a little bit: or rather pushed it out with a little of the inspiration of what is
immortal in someone who has gone before you."

T.E. Lawrence in a letter to his mother, September 1910.

Acknowledgement

She is at once the rock
That founds my weight
That underpins
My sedimentary state

Yet too she is the stone
To which I'm tied
From floating upward
When my mind is skyed

If I should blessed be
To come again
I would that nature set me
Just the same

Contents

Contents

Introduction to Volume One

The first of four volumes, *Volume One* comprises two parts. *Part I – Context* addresses those domains that precondition all else. *Part II – Relations* addresses the path of evolutionary developments, ostensibly from nothing to something, to the advent of organic life, to human life in particular and on to a primordial ability to 'see the light'.

This series is a means of exploring the mind. Not a very good one because of the constraints that language and recording place upon the expression of free thought, but a start, as a peep under the lid of an ordinary man.

As to whether these adventures of the mind fall into the category of fiction or non-fiction we shall leave to those who find it useful to classify books. It may, however, be more comforting simply to accord the status 'non-fiction' to those ideas that correspond to the beliefs of the reader while reserving the term 'fiction' for those that do not – always subject to a change of mind.

In the interest of engaging in a rational search for ideal forms the general theses underlying this work are governed by enquiry and interpretation concerning how the universe operates.

In deciding not to include bibliographic endnotes the author took into consideration the uncertainty of many origins, the transient nature of authority, the elusive character of absolute knowledge, the importance attached to ideas that should stand on their own merits and the essential needs of the reader to draw conclusions with an open mind.

Subsequent volumes two, three and four address the *Intellect, Volition* and the *Affections.*

The Beginning

Man has been asking the question 'what is it?' for as long as he has been able to bring his mind to question. 'It' is of course a linguistic concept that we call a pronoun and use to represent any concept that has no proper name or noun, in substitution for any and all nouns, to identify any combination of any and all such nouns and concepts or in reference to a totality. 'It' is all things to all people yet has no specific meaning except as suggested by the context within which the term is used. In so being, 'it' conveys the secular equivalence of omnipresent existence – a word in a world connoting a world in a word.

It is not possible to trace the relation of everything to everything else for reasons that readily become apparent, but to recognize the implications of such interrelations is to entertain a prospect from which it is impossible to disengage one's own conduct. The pursuit of the holistic view, while not literally an achievable end, is an eminently worthy and thoroughly absorbing undertaking; a claim that resonates well with Plato's persuasion that philosophy is 'the spectator of all time and all existence'.

The questions that stand paramount in being human are what could and should a human being be? Significantly, the most intractable problems of the future will, as in the past, arise from the nature of human mentality and, that being so, it behoves us to gain a fuller understanding of its nuances. What greater challenge has mankind than to master its own mentality?

Critical urgency attaches to realization that, with potential to annul the products of 4 billion years of evolutionary life on earth, mankind has in effect imposed a system of limited sanctions upon nature, and because that is our conscious choice, that is our destiny.

At birth the mind is an empty reservoir. What is the context in which we proceed to draw knowledge if not by making comparisons with the semblance of a full reservoir, one may ask?

Yet, ironically, there is not even a rudimentary model out there. 'It' simply doesn't exist! Being confused by the extraordinarily complex environment in which we find ourselves, the best that the ordinary man can do is to seek reasonable explanations to the questions what, where, why, when and how.

The first thing to be asked of a dynamic system is 'what does it do?', to which the answer has to be 'it works'; confirming the virtuous reality of a world where all is in service to all else, even when we struggle so hard to convince ourselves otherwise! All there is for mankind to understand is its appropriate relationship to all there is.

We establish our values in the light of personal experience and through making comparisons with those of our contemporaries and forebears that come to our attention on the evidence of their behaviour. We gain such insights on a subject-by-subject basis, often through chance acquaintance. We have no comprehensive means by which to compare our own mental lives with those of others, nor to verify the veracity of what we infer from the behaviour of others. In this regard each of us is very much alone. Alone, we necessarily indulge illusions that take the form of a series of reflections; trying to rationalize, i.e. make sense of, life's experience.

On a journey through time, space, energy, matter, relations, the intellect, volition and the affections, the author searches for clarification of the single most critical issue of our times, and of all time: the nature of human mentality. The luxury of thinking and writing about anything that came to mind, came to mind. The mind was to assume the occupation of minding its own business. In the context of such soliloquies, in recognition of the potential value of an orderly frame of reference and in response to perceptions of impending peril, the semblance of a 'Model-A' mind-on-paper slowly emerged – with all its human foibles.

Insofar as experience is illogical, it is incomprehensible. The form of this work therefore takes logic as the guiding principle and lays out the subject matter sequentially so that discussion of one subject leads, through natural relations, to the next. To this end a debt is owed to the work of Peter Roget who, in the structure of his *Thesaurus of English Words and Phrases*, first published in 1852, recognized that the classification of ideas is the true basis on which words, which are their symbols, should be classified. So the form and contents of this treatise have been conditioned and afford a modest basis for mind-to-mind comparison, albeit to be found wanting in so many important respects.

The value of the 'big picture' lies in our ability to view the present as a single frame in a movie that runs for millennia. It is the privilege of each generation to enquire, discover and reveal nuances in that ever-changing picture as circumstances allow, through which processes we may seek to redress the compounding problems that arise from being only human. Such were the aspirations of a rational idealist.

Putting a human mind on paper is a little like parading down Fifth Avenue; not just naked, but inside out. That 'inside' is a perception of what is 'outside', and that is 'it'.

Gary
25 October 2006

Part I

CONTEXT

1   Time: The Quintessential Existential Precondition

God, the Buddha, Mohammed and Einstein met for a quiet lunch at the Pebble Beach Golf Club. The purpose of their meeting was to determine the ultimate parameter of life for lesser beings.

Einstein announced that this was serious business and the subject of a lifetime's thoughtful work. God, the Buddha and Mohammed granted him a long life, at the end of which Einstein proclaimed 'It all depends'. He died in Princeton, New Jersey, on the 18th April 1955 A.D.

Mohammed, concerned that the subject of this meeting was very broad, said that he needed a basis upon which to compare the breadth of its implications. God and the Buddha placed Mohammed in the middle of a very broad desert where he made many devoted friends to whom he declared 'Follow me'. He died in Al-Medinah on the 8th June 632 A.D.

The Buddha thought silently for a long time. Einstein had opted for length, Mohammed for breadth. What was really needed, so he said, was depth. Whereupon God appointed him the defender of the third dimension and conveyed him to a high plateau in the Himalayan mountains. The Buddha believed, was believed and is believed to have contemplated for the rest of his life. He died in India in the year 483 B.C.

God, reflecting quietly to himself, resolved that he needed sufficient time to create an infinite number of shiny little white golf balls. As far as can be seen he is still hard at work.

Such is the inscrutability of time.

God had found this format of informal dialogue to be eminently worthwhile as a means by which to cross-pollinate and fertilize his big idea: life. It had been for this reason that he decided to convene a continuum and to invite anyone who had ever been, is, or will be. Through this device many actors contribute in time to a collective understanding of many subjects. And so it proceeds, has been recorded in the minutes of history, bears upon all present and portends for the future.

Time Consciousness

Human consciousness embraces an awareness of time, time consciousness, whereby events appear to take place in a state of flow, either as time flows past the observer or as the observer moves through time, the difference being indistinguishable. Time appears to envelop all other experiences in such a way that it takes on the character of space to the degree that it extends its influence infinitely in all directions. Having disposed of length, breadth, depth and infinity, one might ask what can be understood about the quality of flow? If time flows forward, why not backwards, laterally or on any or all axes? What is the speed of time in such a flowing condition?

While we measure events in man-made units of time, we cannot gauge time by reference to itself in hours per hour. Does time therefore comprise a rate of change with respect to something else, and if so, to what? Is time absolute or relative? Are future events on-line with the past, merely announcing themselves as present when 'now' flows past an observer? What gives authenticity to time, mere consciousness of time? Concerning the relation between time and consciousness the most contentious consideration is the extent to which one depends upon the other. The key question in this debate hinges on the linguistic notion that time is differentiated by consciousness into three distinct classes of experience: past, present and future. In the absence of consciousness that distinction dissolves. At issue is whether subjectivity is an essential factor underpinning time. On the surface it would appear not to be so. If time flows past an observer, then it also flows past the immediate environment of the observer: mountains, lakes and forests. If the observer is no longer present to observe, it is illogical to assume that time stops or departs with the observer, the more so if a second observer remains in place.

The traditional common-sense view of time has been that if the time span between two events is measured separately by two observers using reliable means, the results can be expected to correspond.

From the familiar equation 'speed = distance x time' we interpolate that 'time = speed/distance'. Speed of what? Distance from where to where? For the purpose of this discussion, and in order to reduce the number of variables, we shall 'assume' that speed is relative to an absolute, constant and maximum possible speed, and that light travels in a vacuum at such speed. With agreement on the 'assumed' speed of light, and disagreement on the distance from a light source to an observer due to movement, the time of transmission necessarily remains uncertain.

The two streams of philosophy on the subject of time, the absolutist view of time as being wholly independent of the universe, held by Isaac Newton and his successors, and the opposing relationist view necessitated by uncertainty, both contend that Albert Einstein's theory of relativity exonerates their positions. Both of the above theories, it can be argued, are mind-dependent and therefore cannot be verified objectively.

If the past is presents that have happened
And the present is a future left behind
Then the future of the past is in the present
And the present of a future in the mind

From these perspectives one draws the tentative conclusion that time is necessarily absolutely indefinite, but absolutely defined by necessity.

Early Perceptions Regarding the Nature of Time

If we look to history to found our understanding of time we will discover a multitude of conflicting concepts, none of which satisfactorily serves to bring unity to our knowledge, merely convenience. The Greek philosopher Parmenides argued in the sixth and fifth centuries B.C. that change was illogical, that logic, not experience, governed actuality, holding that observed events supposedly occurring on a time line were in fact illusions acting on the stationary stage of reality. How, one might ask, can it be argued today that Parmenides was correct?

Consider aggregating all time, space, matter and energy into a unity; let us say a box. Granted everything in the box is moving relative to everything else and therefore to the box, and granted that the dimensions of the box may be infinite, but from Parmenides' frame of reference, the box is stationary and everything within it is at play, acting up, never still enough to be defined as anything in particular; indeed everything, as distinguished from anything, is phenomenal. The totality is phenomenal even though composed of parts that may be tangible by our definition. Time and uncertainty are inextricably wrapped together.

Heraclitus of Ephesus held time to be of the essence of reality, from which one can infer that time preceded all other existences, including God, that necessarily rely upon duration.

Plato, whose philosophic influence still bears strongly upon us today, espoused a difference of degree as to the extent that events were illusions, claiming reality to comprise ideal prototypes, while phenomena as experienced were ephemeral copies. To Plato a tree was a conceptual model which all experienced trees approximately resembled, but not without individual distinctions or departures from faithful replications of the prototype.

Hindus, pre-Christian Greeks, the Chinese and Aztecs viewed time as cyclical, whereas the Christian religion perceives time as a one-way flow. If both views are held to have merit, then their relation and the resolution of their apparent disparity can perhaps be expressed geometrically as a 'helix' whereby cyclical time is relative, for example to the seasons on earth, to life cycles, to realignments of celestial bodies, clocks, people, etc., that is, as measurable calibrations of time; while the general flow of time is absolute or linear, moving along the alignment of the axis of the helix. Thus, by analogy and from particular points of view, at the micro-scale time can be seen to be cyclical and therefore multi-directional, whereas on the macro-scale the general flow of time can be seen to be linear and one-way. The Anglican bishop George Berkeley closed the cycle of speculative ideas concerning the nature of time by returning to the theme of time being an illusion as expounded by

Parmenides and Plato about 2,000 years earlier. The 'where' of time appears to be physically everywhere, but conceptually limited by the point of view and the imagination.

The Beginning of Time

The great religious traditions and cosmologies of the eastern Mediterranean asserted the necessity for first cause in order to explain the existence of the universe. In the absence of a plausible rationale, first cause was ascribed to divine providence and recorded as such in the book of Genesis.

In their early quests for a rational explanation to their environments, Aristotle and other early Greek philosophers were disinclined to accept metaphysical wisdom as rote, preferring to adopt the position that the universe, as its name suggests to us today, was occurring everywhere and at all times, that is, forever and therefore without beginning.

Saint Augustine allegedly was asked what God did before he created the universe, to which he replied that time was an intrinsic property of the universe and therefore did not exist before God created it. This position gains credence from a logic that ascribes to time the distancing of events. In the absence of a universe God was alone. Without events and relations, there would be no time. Thus, the argument would run, God's existence doesn't require duration. Saint Augustine, however, accepted a date of creation at about 5000 B.C., which corresponds quite closely to our current understanding of the dating of the beginnings of civilization, if by such term one marks such beginning as that time when man converted from the hunter-gatherer to the agricultural settler, which occurred at abou. 6000 B.C. in Mesopotamia.

In 1781, in his *Critique of Pure Reason,* Immanuel Kant concluded that the arguments for and against universal and temporal beginnings were equally compelling, which conditions he cited as antinomies or contradictions between opposing laws.

In 1929 a significant boost was indirectly afforded to the commencement school of thought through the discovery, by the astronomer Edwin Hubble, that stellar bodies outside our own galaxy, the Milky Way, were diverging at velocities that increased with distance. In other words, the universe was expanding. The effect of such a finding was to offer two important perspectives on time; the first, that the further one could look out at objects in space the further one was in fact able to look back in time since the speed of light had been established as constant; and the second vision was one that arises by inference, that by reversing the directions and travel times of such astral images one can project periods when they were closer together, or, as it appears, diverged from a single point from which the event subsequently dubbed The Big Bang issued between 10,000 and 20,000 million years ago. At such an event in space and time the universe is said to have been concentrated at a point of infinite density where notions of space and time within the universe would be meaningless. From this we conclude that there was a beginning of time, universal time, before which there were, logically, no events.

The alternative, that the universe is without bounds in space and time demands a redefinition of 'creator' whereby God was either 'creator' by virtue of his eternal relation to other existences and their outflows, or that He was created as all other products of evolution were, after which He could be attributed responsibility for other 'creations' including life forms; a rather humble status for a god but one that we will explore further later. By this rationale the theories of man appear to prescribe 'what', 'where', 'why', 'when', and 'how' God may be credited to have created, arguably defining if not creating a necessity for God.

The End of Time

In the Western view, historiography, the practice of recording successive events in time, is accelerating. This is not to say that time itself it speeding up but that events accorded significance are becoming progressively more numerous within any standard period of time. This, it may be argued, is simply an outcome of the fact of growing populations, their higher rates of literacy, and the burgeoning products of their investigations that range from early agricultural development through the industrial revolution to modern micro-science, electronics and space exploration.

Nevertheless, whether time or human activity is accelerating, when the two are related, our perception of the effect is the same; if the same, then time can indeed be represented as quickening. Historically, this phenomenon has been prophesied and is the basis of the Judeo-Christian concept of consummate time that is said eventually and inevitably to lead to an apocalyptic end. An apocalypse has some basis in scientific thought whereby the expanding universe expends its energy (in about 270 million million years) and collapses in upon itself, reversing the tide of events caused by the Big Bang and concluding with the so-called Big Crunch, culminating at a physical point of infinite density again, at which point in time, universal time may be said to have ended. Conversely, should universal expansion continue indefinitely, all parts would become so distant from all other parts as to render a state of insignificance to their relations or to their subscription to a universal system.

The Significance of Sequence

A simple way to recognize the significance of sequence is to observe the effects of simple mathematical operations. In the sequences 1+2+3, 1+3+2, 2+1+3, 2+3+1, 3+1+2 and 3+2+1 the product in each case equals 6. Each equation yields the same result regardless of the order in which the whole numbers (integers) are introduced. Substituting negative, multiplicative or divisive operations for the additive operation used in the above example yields a similar consistency of results provided that: a) the starting point, an unstated 0, is the same in any series of equations; b) that multiple integers are used; and c) that a single form of operation (addition, subtraction, multiplication or division) is utilized in each series. In other words, there appears to be a preordination, fate or destiny governing the outcomes of such equations, independent of sequence.

The converse does not hold however; that is, where you use the same values for integers but use multiple forms of operation, the sequence of such operations will affect the products of otherwise similar equations, except where operations are self-cancelling. For example: $0 + 1 - 2 \times 3 \div 4 = -0.75$; $0 - 2 \times 3 \div 4 + 1 = -0.50$; $0 \times 3 \div 4 + 1 - 2 = -1.00$ and $0 \div 4 + 1 - 2 \times 3 = -3.00$ are compared. Exceptions to this rule occur where operations are self-cancelling, as when $4 + 4 - 4 = 4$ and $4 - 4 + 4 = 4$, or $4 \div 4 \times 4 = 4$ and $4 \times 4 \div 4 = 4$, all of which obviate the effects of sequence.

One must therefore ask oneself what is the equation 'doing'? If there is uniformity of 'doing', i.e. constant application of the same operation, then there is no change in outcomes attributable to sequence. If there is a change in 'doing', in the mathematical model or any other operation, then there will be a change in outcomes accountable to sequence.

What, one might ask, has this to do with time? Sequence in time is timing, that is, order of succession. So it can be inferred that, according to the sequence of events, subsequent events are made possible or denied. Thus, had Hector slain Achilles before the converse event could occur, Troy may not have been sacked, Helen not liberated and returned to Sparta, and the mix of relative influence of Greece and Asia Minor upon each other may have been weighted more favourably for the latter. Such eventuality would have drastically altered Homer's *Iliad* and *The Odyssey*, and Virgil's *Aeneid*, potentially diminished the authority of the Hellenic period of Greek culture, and conceivably forestalled the development of ancient Greek democracy as we understand it today, upon which Western philosophy has been founded for 2,000 years. Operations are forces in action. History is the record of subsequence.

Chronology: The Fragmentation of Time

The Paradox of Extension of Zeno of Elea in the fourth century B.C. raised the question as to how a finite time period can comprise the sum of an infinite number of dimensionless instances. A further paradox arises in that, while we subdivide time into increments for convenience, we have no clear understanding of the form of the entity that we are so dividing. Measuring spans of time presents no difficulties, a unitary scheme suffices, but measuring a building block does not tell us very much about the body of which it is a part. While there is no generally acceptable definition of what time is, we have developed working hypotheses that are sufficiently useful to enable us to secure benefits in our daily lives. The situation is analogous to cutting a line of infinite length into measurable segments but allowing the end segments to remain infinitely long, a strange mix of the finite and infinite occurring on the same scale. Event sequencing serves to define finity within infinity.

It has been suggested that the human sense of time may in some way be governed by or dependent upon electrical impulses in the brain. This being so, any system of time intervals would necessarily need to be congruent with the frequencies of such electrical impulses. The metaphoric 'biological clocks' of plants and animals are calibrated by their growth and reproduction on seasonal cycles of which they are unaware, all conditioned by temperature.

If one is to maintain the position that time is a subjective interpolation of experience, then one must also accept a multiplicity of subjects and admit multiple times. Chronology then becomes simply a social agreement that 'clock time' is a time-sharing mechanism for convenience; a utility manufactured to mediate between otherwise independent temporal systems.

How can we best describe time in such a way that we relate it to common experience, and calibrate it so that when divided each divisor conveniently remains a useful whole number? The length of a day is the result of a coincidence, from when the sun reaches its zenith in the sky to when it next does so. It is familiar to us all. This we can represent by a circular scale we may call a clock, resembling a map of the northern hemisphere with the North Pole at its centre, to be calibrated along its circumference, around which the sun appears to rotate, as around a sundial.

The ideal system of calibration would be one in which a full measure would comprise the sum of its parts. Such numbers are known as *perfect numbers* because they have the intrinsic, magical quality of equalling the sum of their factors. The number 6 is *perfect* by virtue of its factored parts 1, 2 and 3 adding up to 6. While 6 does not offer many whole-number factors, multiplying 6 by 10 to give 60 does; a device that increases the quantity of whole-number factors to twelve (1, 2, 3, 4, 5, 6, 10, 12, 15, 20, 30 and 60). Using factors as multipliers or divisors of this single and simple number 60 we can now calibrate the earth into 360 degrees of longitude approximately corresponding to the number of days in a solar year; the year into twelve months; the month into 30 days; the day into multiples of twelve hours; the hour (and degree) into 60 minutes and each minute into 60 seconds – which system is a residual of Babylonian astronomy. We've got it! Or have we? Three hundred and sixty days in a year? Well not really, this is a working model! Many lending institutions compute their interest payments based upon the 30-day month and the 360-day model-year for the simplicity that it brings to computation. One might think that the calendar year must surely be one of Nature's securest calibrations of time. Not so! Three and a half billion years ago earth's day was a mere fourteen hours measured in present time. Your homework is to determine when the 360-day year was or will be reality.

The twelve hours refer to daylight hours that an agrarian culture and economy are concerned with, the other twelve being out of sight on the dark side of the earth. Those who are concerned about nighttime play the twelve hours over again or wear a 24-hour watch, those who are not sleep it off. Each hour equates with 15 degrees of longitude so that it all works out at the end of the day. Time therefore depends upon what spin is put upon it.

It is the seven-day week that is the heretic in this otherwise beautiful system; a self-inflicted wound wrought in biblical time. The four-on, two-off, six-day week would work just fine. Are we not time-travelling in that direction anyway? Is it time for another revolution? Neither seven (days per week) nor 52 (weeks per year) is a factor of the 360-day or 365-day year. Stuck with 365 days in the Gregorian calendar we are obliged to add on one day every four years as if to leap over our accounting deficit, the deficit accruing due to the lack of correspondence between the solar year and the number of full rotations of the earth in days per year.

Presentiments

In exploring the relation between temporal and energetic uncertainties, it has been suggested that there is a minimal quantum of time that implies a unitary scheme whereby all time spans comprise multiple units of the smallest building block of time, so-called chronons. This concept earlier applied to matter did not hold, the atom having been divided, subdivided and subsequently sub-subdivided.

Consider for a moment the consequence of the absence of time. We could not differentiate between events! Simultaneity would be all that we had. Isn't this disturbingly close to real experience? Isn't the present all that we have? Some argue that time does not exist until it becomes the present. In this sense there is a beginning and an end of time and it is now, and now, and now. What is real is now. While the appearance of something presupposes its existence at the time of observation, in the case of the star it may be light-years distant and may not exist at all at the time of its observation. Everything that we observe is subject to the same precondition, the speed of light, such that all experience belongs in the past; we cannot experience the present except in a time-delayed format, i.e. an historical context that we improperly call the present, which fact argues in favour of the present being a theoretical point. Our memories of the past and anticipation of the future are merely conjectures, diminishing in probability the further events are removed in time from 'now'. The very thing upon which we most depend for any experience, time, cannot be defined in such a way that it reconciles all of our conceptual perspectives of what time seems to be. With this credential, all experience becomes tainted and all knowledge suspect in an absolute sense.

In order to survive and flourish we need to distinguish and articulate the properties of the environment relative to ourselves in order to apply our energies and resources to good advantage. Therefore, with regard to time, it befits us to subdivide our time efficiently, to allocate time to tasks according to priorities, and to get those priorities in the most advantageous sequence.

The Authenticity of Time

There are two schools of philosophical thought regarding the nature of time: The 'process' school holds that the flowing nature of time is a metaphysical phenomenon and cannot be comprehended rationally. The 'manifold' school maintains that the analogy of time flowing is illusory; that time is a construct of language whereby words like future, present and past are used to reference events relative to the occasion of their utterance. While the 'processors' distinguish future from past as open from closed, the future being indeterminate whereas the past is assured, the 'manifolders' maintain the future to be as certain as the past, not in the predeterminate sense, but by viewing present events as what future events were.

The cyclical and linear ways of viewing time are simply man-made patterns of events or phenomena that share similarities but that differ distinctly by virtue of their occurrences being sequential. The intervals between such events give rise to patterns of time which can be metaphorically described geometrically, as waves, cycles, straight lines, in-line or any other convenient graphic representation of such experiences, solely for the purpose of communicating ideas. Each of us appears to be in our own space and time capsule.

The laws of nature have been assumed to be time-symmetrical. Under this assumption a reversal of time would yield a reversal of events such that broken cups would jump from the floor to the table and reconstitute themselves. This would allegedly contravene the second law of thermodynamics that states that the degree of disorder (entropy) in a closed system always increases, from which one must conclude that, if time were to reverse itself, either the second law of thermodynamics would be incorrect or that the laws of physics would necessarily change with a reversal of time.

Since we are aware, through memory, of events of the past and unaware of future events; these impressions of themselves tend to suggest if not confirm the unidirectional nature of time, whether time is so or not. The cause and effect relation also tends to verify this reading of time but does not deny conjecture and intervention as to causes of events and therefore consequences. In planning we may be construed to be theoretically reversing time sequence since we first address the most desirable future in order that we may then induce the appropriate causes to achieve the desired effects.

Utilitarian Time

Mankind has, from his earliest beginnings, attempted to represent his experiences objectively. Processes of observation, investigation, discovery and verification have enabled such impressions to be clarified and the representation of findings more accurately registered. This has been what we like to call a rational process, justified as enhancements to our ability to secure our own interests.

We have in this manner and over time 'invented' the laws of nature. They are working hypotheses, practical bases from which we can, with a high degree of probability, expect to proceed. Time itself has not escaped such scrutiny. Isaac Newton, acting to enhance his ability to secure his own scientific interests in advancing an understanding of mechanics, distinguished between what he called absolute time, an abstract model of convenience, and other notions of time that appeared to be conditional and variously called relative, apparent or common time.

Newton was simply doing what his forebears had done, acting upon the evidence peculiar to his circumstances and making his own rules of convenience. The fact that others also found his rules to be convenient assured him of a place in their history books.

In the absence of a definitive exposition as to what time is, we focus upon the aspects of time that are useful: the relation of time to the realms of our physical and mental experience. When existing expressions of time are not readily applicable to our current needs, we conceive of new ones. Thus, when a need arose to equate time with the multi-directional quality of space, we 'imagined' and applied such a multi-directional form of time, calling it 'imaginary time', which concept helped in the reconciliation of gravity with quantum mechanics.

Today we have Atomic Time, Barycentric Dynamical Time, Coordinated Universal Time, Dynamical Time, Ephemeris Time, Greenwich Mean Time, Hypertime, Imaginary Time, Pulsar Time, Radiometric Time, Rotational Time, Sidereal Time, Solar Time, Standard Time, Terrestrial Dynamical Time and Universal Time; enough to convey the idea that there is no general consensus among physicists as to what time is, but a fervent desire to capture its essence and harness it to advantage.

The dilemma that time poses is that, not withstanding time being the very essence of existence; we are unable to define a single unitary system of time to our satisfaction. Thus, time appears to be a holistic relational environment, the clock a means of describing it. The true nature of time is that it has the properties that one gives to it until one finds more fitting substitutes for the moment at hand. With such illusory beginnings, how can we hope to understand anything else of which time is such an essential part?

The Value of Time

Each of our lives is finite, if not definite in duration. As such it can be subdivided retrospectively into a finite number of hours and minutes. From this it can be inferred that to waste an hour or a minute is analogous to wasting a small fraction of one's life. The consolation to thinking in these terms resides in our understanding that the value of each hour or minute lived is not equal, provided we meet the minimal and necessary test of survival for each hour or minute so that we can live succeeding hours and minutes. The value of any single hour or minute resides in its potential for being applied to good purpose, which is contingent upon opportunity, ability, will and discipline. While the value of 'life' cannot be equated on the same scale as the value of 'time', perceptions of the value of one are inextricably bound to the value of the other as 'lifetime'.

While time is the quintessential existential precondition necessary to assure duration in the absolute sense, it is time in the relative sense that engages our attention as the primary article of faith by which we relate to the world of multiple existences where time 'counts' – even if it doesn't!

Having examined in general 'the ultimate parameter of life for lesser beings', we shall proceed to examine some of those existences that are permissible under its governance.

2 Space: The Doctrine of Geometry

Imagine being posed perhaps the most outrageous of all questions concerning your own existence. Then imagine that there may be an answer to that question supported by a rationale that is so compelling that you are inexorably led to the conclusion that you do not exist. Such questions we might call fundamental. They are seldom asked and answers to them are therefore seldom forthcoming. The answers are believed to be self-evident, axiomatic and requiring no proof. We are going to acquaint ourselves with similar questions presented to our forebears, the answers to which challenge axioms, rely upon evidence and, in so doing, command credible burdens of proof, hopefully bringing us closer to more satisfactory ways of responding to such seemingly naive but troublesome questions.

The Concept of Dimension

'In the beginning God created the heaven and the earth.' So commences the most popular book ever written. Beginnings occur with the initiation of the first in a succession of events, so-called 'points' of beginning. Pythagoras, the Greek philosopher of the sixth century B.C., perceived the point as the smallest unit of volume, much as the numeral 1 was the smallest unit of number, lending support to the idea that matter comprised aggregations of points.

This volumetric concept of the point led directly to the practice of representing ratios using points in matrix at equal intervals, a system that exhibited a lucid means by which to express both the formal nature of the ratio and the numerical value of the components of the ratio visually and without numeration. Thus, a 3:4 ratio was expressed as a rectangle described by three equally spaced points in one direction and four perpendicular, each point, not the spaces between them, representing a single unit of measurement. Aristotle, in the fourth century B.C., asked his peers to accept that the extremities of the end points defined the extremities of the straight line.

On the present-day premise that any quantity is mathematically divisible, Pythagoras' smallest unit of volume, the point, may be divided into two smaller points such that each part is half the volume of the original point. The division may be repeated dividing the two points into four, and so on, infinitely. At the original scale Pythagoras' point is visible, at smaller scales, however, a stretch of the imagination is required to believe in its physical existence, allowing for subjective variations in acuity, until no reasonable person would claim to be able to see such a subdivision of the original point. It has passed from the realm of being capable of physical verification by the senses to become a theoretical construct in the realm of the mind. Conceptually it is difficult to imagine two nothings connected by a something. The idea is neither logical nor fantastic, but rather a kind of marriage of the two. Yet this is what we ask our peers to accept: that two abstract, ethereal, vacuous, theoretical points define the extremities of a straight line.

Now take that line from nothing to nothing, of finite length but infinite thinness, and rotate it so that you view the line in-line with (radiating from) the left eye, closing the right eye. What do you see? A theoretical point; nothing! It is impossible from this vantage to entertain the idea that you are looking at something with dimension. However, you are looking at a zero-dimensional view of a one-dimensional line; Aristotle's point! If you enlarge that infinitesimally thin line to perceptible thickness, say a rod, and open your right eye, what do you see? You see a definitive thin line of finite length, in perspective. It is as if one needs to open additional eyes to see additional points of view or dimensions. Without such additional points of view the imagination is taxed to entertain a fuller understanding of the true nature of a viewed object.

Single-Dimensional Linear Geometry

By Pythagorean logic, a series of abutting points on the same straight alignment would form a line of known length determined by the number of points used to construct the line. By Aristotelian logic, only two points would be needed to construct the same straight line. Insofar as a theoretical straight line has no width or depth, but only length, it can be seen to be uni-dimensional. The only true measurement that can be applied is that parallel to its axial length. Similarly, a curved line, including a closed loop as is the case with a circle, must be acknowledged to be one-dimensional if, like a straight line, it can only be divided into parts along its longitudinal axis. Thus, a line, distinguished from the area it may enclose, may be said to be a construct in single-dimensional geometry.

In about 300 B.C. Euclid, a young Greek mathematician teaching in Alexandria, was to successfully challenge the laissez-faire beliefs of the pre-scientists and to lay the foundations to spatial understanding for the next two thousand years. Drawing upon precedence and exercising the deductive methods utilized by Pythagoras, Euclid applied the same principles to mathematics that Socrates had applied to dialectics nearly 200 years earlier: a step-by-step logic intended to pursuade the listener of the efficacy of a complex idea by breaking it down into simpler, more comprehensible parts, and soliciting acceptance in increments. Euclid's fundamental contribution was to espouse the necessity for recognizing that the broad acceptance of mathematical processes depended upon the application of rules. His thirteen-book *Elements* propounded the rules of mathematics known to his time as garnered from previous authorities, substantially rearranged and augmented by Euclid. Euclid's *Elements* is said to have enjoyed a greater circulation than any book in history except the Bible.

In the *Elements*, Euclid introduces fundamental geometric concepts: the point, line, plane and angle, sometimes without definition, and further introduces propositions regarding these concepts, which he asks the reader to accept as being 'axiomatically' correct based upon the relation of such concepts to our experience of the physical world. He then proceeds to define special terms that utilize these geometric concepts, each having a limited and specific meaning, and to coin names to describe them: the triangle, the right angle and the hypotenuse. Accepting the above 'axioms', Euclid introduced geometric theorems of which the theorem of Pythagoras, stating that the square of the hypotenuse of a right-angled triangle is equal to the sum of the squares on the other two sides, is an example.

The premise underlying Euclid's geometric theorems was that – given certain information about a geometric figure, and a statement of intention to prove the correctness of a supposition – then other information can be logically deduced that will be sufficient to confirm the correctness of the original supposition. A practical consequence of the use of Euclid's theorems was that an audience could be led from acceptable assumptions, through a series of simple steps, to accept proof of a proposition that it would not likely have accepted at the outset had it been presented with that same proposition in the form of an assertion.

Euclid was concerned with establishing the positions of elements of rigid bodies relative to each other. To this end he employed a singular concept, distance, to describe relative positions such that any distance can be described as a ratio of any other distance, equality arising out of coincidental sameness confirmed by repeated experience. 'Distance' may be considered synonymous with 'straight line'.

The next step was to seek a unitary basis for describing distance by building upon the metric point system of Pythagoras. The concept of measurement, the assessment of magnitude of a distance using digital units, had been utilized by the Babylonians. The Babylonian basic units of linear measurement were the cubit, approximately 528 mm (20.9 inches), and the foot, two-thirds of the cubit. They were deduced from archaeological evidence and derived, as were other early systems, from the length of human body parts, the cubit being the length of the forearm from the elbow to the end of the middle finger. Clearly such a system was imprecise but apparently sufficient for their purposes. The Egyptians used a royal cubit of a slightly shorter length, 524 mm (20.62 inches), and the foot, again approximately two-thirds of the cubit, as units of measure, subject to the same limitations.

The early Greek basic unit of linear measurement was the finger, sixteen fingers describing one foot. While this unit of measurement may give anatomists a moment of consternation, the length of the human foot does lend itself to division into sixteen equal parts that approximate the width of one finger 19.3 mm (0.8 inches), the origin of such logic. It is noteworthy that today, colloquially, the finger is still in use in the Balkan countries for measurement in the service of alcoholic spirits. The Greeks established separate descriptors for serialized multiples of the finger width, rationalizing them in the sequence 2 (the knuckle), 4 (the palm), 8, 10, 12, 16 (the foot), 18, 20, 24 (the cubit), 40, 72, 96 and 100 (the pole), thus inferring a decimetric system that was to await redefinition by the French clergyman Gabriel Mouton in the seventeenth century.

Two-Dimensional Plane Geometry

With the ability to express distances in terms of multiples of equal digits came the opportunity to substitute abstract symbols to represent complex descriptions. The way was open to apply such abstractions in lieu of words to the solution of theorems. With a unitary measure of distance in hand, symbols could be assigned to identify each element in, for example, a right-angled triangle, whereby the vertex of an angle could be described by a single symbol, say 'B'; the lines defining that point by two symbols, say 'AB' and 'BC'; and the angle subtended by the two lines by three symbols, 'ABC'.

Geometry, named by the ancient Greeks to describe their early practices in earth measurement, was understandably focused initially upon resolving the working difficulties that arose in the measurement of flat surfaces; hence the term 'plane geometry'. With respect to early surveying and mapping, the surface of the earth, albeit curved, could be construed to be two-dimensional because any point on it was identifiable through the use of just two coordinates. Concepts of plane geometry do not necessarily depend upon bodies being in a rigid state, but theoretical planes are not so amenable to testing.

Greek tradition holds that the elements of geometry came from Egypt. Eudemos of Rhodes (fourth century B.C.) maintained that the first geometer was Thales of Miletus, who flourished in the first half of the sixth century B.C. However, subsequent findings in excavations in Babylon, Mesopotamia, appear to open his assertion to question. A more plausible thesis, attributed to Proclus (410–485 A.D.), was that Thales introduced geometry to Greece. Thales' alleged measurement of the Egyptian pyramids and his proximity in Asia Minor to sources of mathematical ideas emanating from Babylon would have enabled the influence upon Greek algebra and astronomy that has since been ascribed to Mesopotamia. The Egyptians had not progressed beyond the practical application and relation of geometric forms: two-dimensional triangles, rectangles and three-dimensional pyramids and spheres. Thales is credited as the author of five geometric theorems on the basis of which a reference by Herodotus (fifth century B.C.) to Thales' prediction of an eclipse of the sun, believed to be that of 28th May 585 B.C., is plausible.

Significance can be attached to Thales' attempts to explain natural phenomena through systematic simplification and deduction. In so doing he anticipated processes of rational analysis later to be employed by Socrates and Euclid. In this regard, Thales may be recognized for pioneering distinctions between myth and reason.

The Greek philosopher Plato, in constituting his Academy in about 387 B.C. for the systematic study of philosophy and science, laid the foundations to rational principles that were to influence thought through succeeding millennia. By Aristotle's account, Plato recognized that there is only one type of right-angled triangle with sides proportioned in the ratio of 3:4:5 that represents an example of 'reality' consistent with Plato's *Doctrine of Forms*. An infinite number of similar triangles with sides of differing lengths but in the same proportions, he would describe as 'unreal' under the same doctrine. The disposition behind such thinking is that there is a singular idea, the model triangle, which is held to be constant and therefore real, but which is illustrated by means of examples by numerous sensible images that are incorporeal. It is not, therefore, entirely irrational to suggest that Plato's view of the world was a construct of numbers and their relations, which, one might argue, could be held to be the key to a greater understanding of all of nature's mysteries; an idea that can be traced to his dialogue *Epinomis*. The conception of the point being the smallest unit of volume, Aristotle suggests, led Plato to deny the notion of a point being strictly theoretical. In accepting whole numbers as the only units of numeration, Plato, like the Pythagoreans, was denied the prospect of resolving more complex problems requiring the application of irrational numbers, including square roots, cube roots and pi (π), thus leading to a decline in the influence of the Pythagoreans and the extension of their theories by others at later dates.

Three-Dimensional Volumetric Geometry

Of scientific and mathematical significance, Plato constructs his world, not with matter but with *chora*, or space, expressed in his dialogue *Timaeus*. The notion that space replaced matter was later to find greater security in the work of Rene Descartes.

Eudoxus of Cnidus introduced new insight to geometry in the fourth-century B.C., devising the 'Method of Exhaustion' (so named in the seventeenth century) by which irregular areas and volumes were approximated by combining the measured volumes of slices through the form, a process still in use in the construction industry for estimating quantities of earth.

Menaechmus, a student of Plato and Eudoxus, is remembered for his attempts to master a challenge of an architectural nature, the so-called 'Delian Problem' of constructing a cube twice as large by volume as a given cube. His efforts to solve this problem led him to investigate and describe the geometric properties of the cone and to recognize that plane intersections of the cone, known as conic sections, depending upon the angle of intersection, generated the circle and curves not known in nature at that time; the parabola, the ellipse and the hyperbola. Building upon the earlier work of Eudoxus, Archimedes greatly extended the 'Method of Exhaustion', achieving levels of sophistication equivalent in some cases to concepts of integration to be rediscovered and propounded 2,000 years later by Johannes Kepler, Bonaventura Cavelieri, Christiaan Huygens, Descartes and others.

The Greeks disfavoured algebra as practised in Babylonia, believing it, in the absence of fractions and decimals, to have serious limitations when applied to commensurables (numbers having common divisors). They developed a clumsy system of transposing algebraic problems into the language of geometry, so-called 'geometric algebra', but could not solve problems more demanding in this way than could be solved using a measuring stick and compasses.

Greek geometers focused their attention upon the pursuit of proofs to their theorems through the exercise of reason rather than upon reconciling their concepts to experience. In this sense their activities can be viewed as intellectual games rather than the pursuit of empirical knowledge. A kind of abstraction prevailed whereby they were intent on objective examination at the expense of subjective interpretation.

Curved Planes

The Treatise on Conics (sections of cones) in eight books by Apollonius of Perga, in the latter half of the third-century B.C., gained for him the title of the Great Geometer for which he is remembered today. The first four books contained the essential principles. Apollonius confirmed the earlier observations of Menaechmus concerning sections of right circular cones; proving the general rule that plane sections of any circular cone, taken parallel to an element, line or plane, described either a circle, a triangle, a parabola, an ellipse or a hyperbola, which terms Apollonius used for the first time. This accomplishment is regarded as the climax of ancient Greek geometry.

It is significant that Apollonius appears to have regarded conic sections as plane surfaces rather than as elements of three-dimensional and therefore spatial forms. This may have been due to his extensive use of different coordinate systems, always in the mode of 'geometric algebra'.

By the year 200 B.C., almost all the classical Greek structures of which we are aware today, in the sophisticated Doric, Ionic and Corinthian orders, had been completed. Yet, ironically, this flourishing and refined art of building had not harnessed and exploited the solid geometric forms of which the ancient Greek geometers were so fond. This, one might conjecture, was due to a combination of the following circumstances: their naivety in treating geometry as mental exercises without recognizing its potential for practical application; technological limitations arising from the nature of the limited materials at hand; and a prevailing political climate that deterred deviation from a strict set of socially acceptable building design standards – all of which require further examination.

Eighteen hundred years were to pass before the seventeenth century was to yield fresh fundamental thoughts concerning theoretical geometry. They came first in the form of a convolution of Euclidean ideas by Descartes.

Coordinate Geometry

Descartes, building upon the earlier work of Archimedes and Apollonius, is credited with the systematic development of 'coordinate geometry', so named by Gottfried Wilhelm Leibniz and also known as 'analytic geometry', by which the position of a point in space may be described by reference to ordered sets of numbers called coordinates. Geographic degrees of longitude and latitude, when combined with elevations above sea level, describe points within the system that were implied but not employed by Euclid. Descartes is reputed to have had a dream revealing a system by which physics could be reduced to geometry, and all the sciences be interrelated, upon which he is said to have based his life's work. Publication, at the age of 42, of his work *Discourse on the Method of Properly Guiding the Reason in the Search for Truth in the Sciences*, in which he rejected as 'a collection of curiosities' indefinite notions such as weight, hardness, lightness and heaviness in favour of geometrical absolutes like the point, plane, straight and circle. In so doing, Descartes laid the foundations to the modern scientific method by which verifiable knowledge could be seen to flow from the application of general principles. Mathematics, he asserted, was to become the common denominator of all scientific endeavours by virtue of its reliance upon, and reduction of, concise definitive data. His *Discourse* set about the task of searching for universally applicable laws of nature that would validate his thesis and replace the then-current multiple and dissociated 'explanations of existence' with a single system of reasoning.

He rationalized that in order to gain insight into complex 'natures' it was necessary to reduce their complexity through analysis, to intuitively identify and distil the fundamental character of each part, and to reconstruct those parts in ways that would enable a clearer understanding of the original complexities utilizing a singular system of recombination. Descartes' 'natures' may be likened to letters of the alphabet, or elementary atomic particles. Limited in number, and capable of assemblage to form vocabularies of words or molecules of known and novel combinations, each combination may in turn be aggregated to compose disparate concordant arrangements conforming to the rules of the discipline to which they belong, as to language or chemistry.

A practical, modern-day analogy would be to recognize that buildings by type, be they civic monuments, houses of worship, railway stations or schools, are combinations of elementary parts: masonry, wood, steel, glass etc. According to Descartes, what was necessary was a clear understanding of the nature of each part, its relation to other parts and the sequence by which the parts needed to be assembled in order to arrive at the intended end condition. He had defined and espoused the scientific method as we know it today.

In this manner, he argued, a reasonable person could 'read' and therefore discover the essential nature of a complexity by virtue of knowing the language and syntax necessary for its decomposition.

Descartes was, however, unable to account for the special properties that attend critical combinations of parts that establish discrete totalities. This was most evident in his attempts to treat biology as he had physics and chemistry, mechanistically. His descriptions of animate bodies as natural machines seemed to expose a flaw in his rationale: failure to recognize and define the special attributes of whole systems, most notably the special natures of life-bearing forces, including human will.

Concerning movement, Descartes observed:

'If a moving body impinging on another has less force to continue moving in a straight line than the other has to resist it, it loses nothing of its movement though it changes its direction; but if the former body has greater force it moves the other body with it, losing so much of its movement as it imparts to that other.'

Through the introduction to geometry of the ideas of constants and variables (enabling curves to be described in algebraic form) and by correlating the properties of sets of numbers and points, Descartes was able to demonstrate that the numbers and points were interchangeable. Geometry could now be expressed in terms of algebra, not a new concept, but conversely, algebra could be expressed in terms of geometry; Descartes' salient contribution to geometry. Through these developments Descartes was able to advance the interests of Menaechmus, mastering the problems of doubling the volume of the cube and trisecting an angle, and in so doing to contribute to the theory of equations.

Elliptical Thinking

Concurrently, the German astrologer Johannes Kepler was absorbed with the idea that there must be some systemic pattern to the relationship of the orbiting planets. He was particularly struck by the eccentric motion of Mars, which did not appear to conform to the then current and 2,000-year long-standing presumption of planetary movement in circular orbits. Taking the bold hypothesis that the planets circumscribed the sun in elliptical paths, Kepler set out to investigate the kind of motion that would result in such elliptical orbits. His extended labours were rewarded in 1618 when he discovered that the radius vectors (straight lines joining the centre of an attracting body, the sun, to the centres of the planets orbiting around it) described for each planet equal areas of the ellipse in equal times, thus confirming the elliptical nature of planetary orbits and revealing for the first time the occurrence in nature of the ellipse, which form had merely been theoretical to Menaechmus and Apollonius.

Isaac Newton's greatest contribution to an understanding of space may be seen to be a by-product of his enquiry into the forces that operate in it. His discovery of the law by which force decreases in intensity with the square of the distance (known as the inverse-square law) established a common connection through observed behaviour between the movements of celestial and earth-bound objects, including the tides, thereby affording a scientific language of spatial relatedness that had hitherto been missing.

Exotic Geometries

Projective geometry was developed to address the special difficulties arising from the need to accommodate perspective in the representation or mapping of one plane upon another. It evolved to encompass geometric phenomena that were more abstractly detached from everyday practical problems. Perspective drawing and similar practical applications are now described under 'descriptive geometry'. Suffice to say, 'projective geometry' addresses the connections between algebra and geometry that led to the formulation of a special branch of mathematics, 'algebraic geometry', at a later date.

Gerard Desargues and Blaise Pascal pioneered the field in the seventeenth century through the formulation of two complementary theorems. The first described in-line points resulting from the projection of corresponding sides of certain paired triangles in the same plane. The second, intersections of lines joining points on a conic section occurring on a straight line.

The discipline has become more precious or esoteric over time. Special conditions have been identified wherein, for example, a line may become perpendicular to itself (isotropic lines); the sum of the angles within triangles exceeds or is less than 180 degrees (in elliptic and hyperbolic geometries respectively); and where (in field theory) $1 + 1$ may equal 0 as occurs in skew fields where A x B is not necessarily equal to B x A.

Of particular relevance to this discussion, any aggregate of 'things' falling into classes and subject to special relations can be expressed to give a particular projective geometry; the 'things' being points grouped into classes of various sizes called lines or planes etc., the number of dimensions to the geometric form depending upon the number of different sizes of classes.

N-Dimensional Non-Euclidean Geometry

The suspicion that dimensionality of space has as much to do with perception as with mathematics led Friedrich Gauss to suggest that space was not inherently defined by a certain number of dimensions but rather that the number of dimensions was 'elective' and depended upon what one was trying to 'prove'. If we are using the term 'space' to describe separation, then we use a single dimension. If we seek 'space' to contain an area, we use two dimensions. If we try to quantify 'space' as volume we use three dimensions. Experience does not therefore confirm or deny a specific and intrinsic number of dimensions that circumscribes space, but rather the need exists to 'select' the number that best 'fits' a particular purpose in describing space. Thus, there is an element of 'faith' that places geometry on a footing with religion, suggesting that 'belief' may be analogous to a kind of 'algebraic equation' that 'works' if it serves to best 'fit' or explain experience. Since common, everyday experience of space is volumetric, Euclidean geometry with its three dimensions remains the popular geometry of 'choice'.

A serious challenge to the classical concept of Euclidean space was extended by the mathematician George Riemann in the nineteenth century. He renounced the idea that freely movable rigid bodies should be used to define three-dimensional space. His substitute proposal: the use of a hypothetical model involving a metric structure of space. Riemann's geometric model is noteworthy for his introduction of the idea of a manifold, a topological space represented by a set of 'local' coordinate systems that are linked to each other by specific conversion principles of a common kind. The appearance of continuity of curvature is explained as being the sum of infinitely small flat parts. A complete manifold so defined is called a space-form. While Riemann departed from the earlier two-dimensional rigid body geometric concepts of space, he substituted, in effect, a characterization of space defined by a one-dimensional system utilizing freely movable rigid rods or bodies, thus disallowing the idea that space was *a priori* given, insisting that it was determined by matter.

When addressing space on a universal scale man is limited in his point of view as an insider looking out. He can only define the inside limits of space. Such was the essence of Riemann's procedure that lends itself to an interpretation of architectural space. Since verticality is merely a local condition unique to any line radiating from the centre of the earth and horizontality is a short chord of the curvature of the earth perpendicular to it, architects are designing in Riemannian geometry and their structures comprise multiple Riemannian manifolds.

Mathematical Self-Evidence

One such divergence concerning the nature of space was that proffered by Hermann Helmholtz. In accepting the traditional premise of the existence of freely movable rigid bodies, he hypothesized that, since distance is established as the relation of pairs of points, and since they could be 'mapped' or reproduced upon themselves, space itself was a rigid body. Was this not an echo of the ancient Greek idea *chora,* substantiated by state-of-the-art mathematics?

The notion that knowledge could be acquired through deduction, as was the case with geometry, made geometry a special exception. Its internal processes appeared to provide its own mathematical evidence, proofs and truths beyond doubt. Interpretation and opinion were thereby disallowed. A passage to reason was revealed.

Reasonable Doubt

However, seeded within the internal processes of geometry were latent soft spots awaiting the challenge of more rigorous minds. The exercise of judgement is one such area of contention. The doorway to geometry lies ajar. Before one may enter to play the game one must judge the existence of *a priori* conditions. Such conditions are generally so elementary in nature that all who enter mathematics are presumed to share knowledge of and concur in their existence.

Geometers call such preconditions postulates or axioms and, being agreed upon at the outset, they are not subject to proof. They pass through the door as unchecked baggage. Nevertheless, in order not to corrupt subsequent play, one might reasonably expect such baggage to pass explicit benignity tests. Axioms are synthetic; they are composed by man; they embody assumptions; they are purported to be *a priori* certain, embodying ideas that do not depend upon each other as determined by analysis. They impact the way the game is to be played with the things and the relations. The above minimal descriptions do not qualify axioms as being error-free and therefore unassailable truths.

According to Moritz Pasch: 'Whenever geometry has to be really deductive, the process of inferring must be independent of the meaning of the geometrical notions as well as of the figures. The only things that matter are the relations between the geometrical notions, such as (have been) established in the used theorems and definitions.'

It can be seen that the forcefulness of geometry as the bulwark of realism was beginning to falter as physical and mental constructs appeared interdependent. This fusion was conclusive with the publication in 1899 of *Grundlagen der Geometrie* by David Hilbert, with the opening words '*Wir denken uns ...*' ('We imagine three kinds of things...'). He is, of course, referring to points, lines and planes and we have returned to the open question as to what is 'real' and what is 'imagined'. Hilbert had broken the sensible space barrier. No longer was geometry to be relegated to merely the parochial role of designator of spatial experience. A fuller understanding of geometry now required extrasensory perception. Notwithstanding geometry's deductive passage to reason, we shall continue to show how distinctions between the 'real' and the 'imagined' are blurred, difficult to draw, or undifferentiable.

No one so eloquently disclosed the importance of this turn of events as Albert Einstein in a 1921 lecture Geometrie und Erfahrung: 'As far as the mathematical theorems refer to reality, they are not sure, and as far as they are sure, they do not refer to reality.'

Geometry as Metaphorical Modelling

Hilbert had recognized a distinction between the axiom that had hitherto been acceptable as a truism, and what an axiom needed to be to satisfy the independence of its integrity. Such a substitute axiom was to function as a model, in Hilbert's case in algebraic form, the test of its independence being its replaceability with alternative models. In this way an algebraic model became an analogue for and the path to the solution of a geometric problem, a reversal of the ancient Greek substitution of geometry for Babylonian algebra referred to earlier.

Compound Reproduction

Geometry is a relational system, the simplest element of which is the point. Abstractly, a single point has no relatives; it is an orphan and has nowhere to go to find affinities. The simplest expression of geometric relation arises when a second point is introduced.

The simplest way to obtain a second point is to mirror image the first. Behold we have geometrical reproduction, or transformation as it is called, in this case symmetry about a theoretical line equidistant from the two points and perpendicular to a straight line joining them. Repetition of such reproduction about parallel lines, known as translation, and about intersecting lines, known as rotation, produces, not surprisingly, translations and rotations respectively (with exceptions). Reproductions assume familial forms analogous of organic growth as the geometry is extended. By such methods groups of forms, with their own characteristics, can be defined. Such are the fundamentals of mapping exemplified in architecture that allow us to define space in ever-increasing complexity through the use of conceptual building blocks rather than the association of individual points, lines and planes.

We now have families of geometries suggestive of Riemannian manifolds where the characteristics of separate families are dissimilar. What is the significance of their mutual discordance to their interfamilial relations? This is the question central to an understanding of all relationships, the answers to which are essential in order to gain access to their combined optimal performance as parts of greater entities. We shall discus this recondite subject in due course.

The Extent of Space

Let us make a clear distinction between 'space' in the abstract and the 'universe' to which we ascribe certain properties. If we accept that space is a continuous environment that fulfils the function of physical containment, and that all phenomena are so contained, the question arises whether space extends beyond such phenomena if and where they do not exist. In other words, do we place limits on space as container as determined by the function of containing? Theories of the universe tend to do so on the premise that if there is nothing beyond, there is nothing beyond to discuss and such limitations are self-prescribed! However, we must entertain the alternative question: what if there are no limits? What if space is constant and invariant? Then there is nothing beyond because there is no beyond. The distinction that we need to make between 'universe' and 'space' is that 'universe', by our definition, connotes 'all-there-is', whereas space contains 'all-there-is'. The question now arises, what is beyond the universe if it has limits? The answer appears to be more 'space', as an extension of our construct 'container' but without necessity for limitation in size. We are all familiar with less than full containers, even if we have a little difficulty imagining a container with limitless capacity – Parmenides' 'box'. We conclude that space is limited only if it is itself constrained to contain.

If, as science argues, the universe has form (shape), then by extension there must be an existence outside that form through which we must entertain the possibility that our universe, while 'uni' or singular, may be one of 'multi'-ple unread 'verses' out there! We find ourselves in a similar quandary to that we confronted in seeking first cause, whereby we must ask ourselves: if one of 'multi', then how many? A question that we have no way of responding to satisfactorily.

One of the stumbling blocks for laymen in trying to understand physicists is that, conceptually there seems to be no reason for placing limitations on the extent of unenclosed space. In contrast, physicists place a limit on space that is synonymous with universal space without so stating, understanding that the laws of physics as we know them may not apply beyond our universe.

If space extends infinitely in all directions then it doesn't matter where the coordinates are, they may extend in any and all directions to infinity also. Thus, everyone may be a Riemann with his own personal spatial coordinate system. It is not necessary to have a universal coordinate system per se. In order to communicate ideas, all that is needed is an agreement to share a particular coordinate system to enable collaboration in the pursuit of particular knowledge. That is just what Euclid offered.

Today the Global Positioning System computes locations in space on the basis of triangulating distances from multiple earth-satellites. Theoretically, a discerning instrument could reference any coordinate system calibrated in any linear system of measurement.

The most logical global coordinate system would be one that references an alignment of absolute zeros; that is to say, zero degrees longitude, zero degrees latitude and the centre of the earth. This base line would surface at sea level in the Atlantic Ocean about 600 miles west of Libreville, Gabon, 400 miles south of Accra, Ghana, and 3,963 miles from the centre of the earth, references that most of us have little interest in most of the time, so we take our bearings from this point by abstraction.

Pandora's Box

We have arrived at a similar predicament in examining 'space' that we reached in discussing 'time': that we do not have a firm fix on the nature of the subject of our interest. Instead, we 'invent' parameters to suit our purposes: Architectural Space, Curved Space, Eleven-Dimensional Space, Euclidean Space, Five-Dimensional Space, Hyperspace, Interstitial Space, Mental Space, Platonic Space, Seven-Dimensional Space, Space-Time and Topological Space – enough to fill Parmenides' infinitely large 'box'! When we address the subject of space it is as if we open Pandora's mythical box and invite a multitude of unforeseeable difficulties. It would appear that any definition of space is acceptable if, when applied to an intended end, it were found to be useful. In other words, we have defined space per se as having no intrinsic qualities at all, save dimensionality. What do these revelations tell us about the states of science and mathematics, our most disciplined search-engines of truth? We suffer from an obdurate confusion between space as volume, dimensions that are means of describing it, and what space contains, which are merely contents.

We have dwelt at some length with geometrical ideas for a number of reasons, the most obvious of which is that the integrity of form relies upon the closing of geometries consistent with definitive principles. More broadly, geometry lends itself to being applied metaphorically to elucidate mathematical, linguistic and visual relationships, as stepping stones on a path through perception to cognition.

Space, it can be argued, must have preceded God for without space God cannot be anywhere or everywhere when all there is, is nowhere. This relegates God at best to third place in the line of creation after time and space. So, in the beginning, time and space accommodated the creation of God!

Finally, it is noteworthy that as concepts of space expanded to put aside previously held limitations, the mind of man may be considered to have broadened to accommodate the new ideas. Whether the ideas expanded the mind or the mind the ideas is a moot point.

3   Energy: Inherent Potential

Perhaps we should start the discussion of this subject by dispelling a little confusion between a few related concepts: energy, force, work and power. Energy is the ability to exert force, do work or produce change by virtue of potential in reserve or the fact of motion. Force is the release and exercise of the potential of energy in a direction, as a push or pull. The mythical 'push-me-pull-you', with a head at each end, is an ambivalent force-horse. Work is the product of the application of force over a distance, and power is the amount of work performed in a measure of time.

Nothing

Imagine an environment where nothing, not a thing, exists. There would be no matter, no energy, no space to relate them in and no time to relate them to. There would be no them, no it, not even a god. This is the environment of all-there-isn't.

Something

If any 'thing' were somehow to enter this environment, it would have to come from without, from another environment, as from one universe to another where crossing from one to another had hitherto been denied. We are now in a position to entertain the idea that there is more than one environment, two, or possibly more. If we now extend our horizons and declare the new combination of the two known environments to be the environment of all-there-is, the first question that comes to mind is where did that thing come from before it moved to the environment of all-there-isn't? There would appear to be but two answers: the first would be that it had always been in the second environment and had somehow escaped and lost its way; the second answer would be that it came from yet another environment to which the same question would apply: where did that thing come from before it moved? We are at a loss to explain first cause. From our perspective, probability suggests that there is no first cause of energy; it has existed, does and will exist, as far as we can tell, merely transforming itself from form to form and place to place from time to time. Of course our perceptions are limited by our experience and imagination. We cannot deny that somewhere there may be an eternal fountain of energy being constantly replenished through the creation of energy from its opposite number, non-energy. We simply haven't met non-energy yet; positive and negative energy yes, but not non-energy or anti-energy. Science has accepted this probability and placed limits on the total quantity of energy in the universe, the only universe we know of. All-there-is, is all there is! A finite quantity of energy is represented by joules, albeit often masquerading as chemical energy in matter. We acknowledge the caveat that the fountain cannot be denied with certainty. Nor can God. Nor can the possibility that the fountain and God are one and the same.

So the thing proceeds in its invasive way, all juiced-up with energy in an environment where there is otherwise none. What is it going to do? It is the sovereign of nowhere-land, and so, like any sovereign – god, king or thing – it is going to radiate its authority and through that radiation distribute its influential parts. A small thing would do its thing modestly, but a large thing would do it with a 'big bang'! The rest is history.

Things

As you woke up this morning the radiant distribution of influential parts continued. In their earliest moments they had appeared to be much alike and were a small family of particles that we know today as photons, electrons and neutrinos, their polar-opposite numbers, and a few protons and neutrons. Very hot stuff! Now most of the electrons (which should have been called vanitrons), being inexorably attracted to their opposite numbers in the shaving mirror, anti-electrons or positrons, broke their mirrors in their excitement, cut themselves and bled to death, a death shared by their reflected images, the remnants of both becoming photons (analogous to photographic memories). But a few electrons survived. A while later, as it became a little less crowded and a little less uncomfortably hot, the particles had a chance to look around. The protons got to talking to the neutrons and struck up affinities, and, with a little help from their friends the electrons, formed nuclei families of the first atoms, heavy stuff, heavy hydrogen or deuterium, named from the Greek word *deuteros* for combination or second generation of form. Man, having an insatiable curiosity, once believed that if he could understand this process step by step, he could trace its path back to the beginning and recreate creation within his own authority. The result was the hydrogen bomb.

Attraction

Since deuterium there has been a steady spawning of little 'things', the essence of each of which has been the establishment of affinities, that is to say attractions to opposite numbers, to which we attribute polarities of force: positive and negative. Which is which and who is what is merely opinion, a point of view, for he finds her attractive and is therefore positive, as she does him and so feels the same, while each also feels such compulsion to be beyond its control, negating its identity, violating its integrity, and therefore negative. The first such mating was, metaphorically, original syn (pronounced 'sin') from the Greek word meaning 'with, along with or together'. The corollary of the attraction between dissimilar forces is the general rule that like forces repel.

What caused positive and negative energy? Absolutely nothing! It was the presence of nothing, upon being introduced to something, that established difference in identity, the opposite number, the foundation of the yin and yang of ancient Chinese cosmological traditions, where the dark, feminine and negative principle yin combines with its affinitas, the bright, masculine, positive principle yang (surely a description planted by man for woman to give issue to), to produce any 'thing' and all-there-is.

Confusion

After breakfast you may have considered what forms the radiating influential parts may have taken. If we think of parts or particles we tend to think of them in a material sense and assume that they move in straight lines; if we think of energy we tend to think of ripples on a pond travelling in waves but, as we shall see downstream, and could see in any stream, waves and particles commingle in such a way that the waves resemble particles and the particles waves, a confusion that energy and matter use to bombard us with in our innocence and their radiating influence. Wave/particle duets may be viewed as tunes of nature struck between energy and matter as their contributions to universal harmony.

Personality

Nature endows all broad classes of particles with personality such that any one can be distinguished from another based upon appearances. This endowment is called 'spin'. As you rotate in your relation to others the appearance of your head is transformed from a full face to a profile, to a rear view, to an opposite profile and back to the full face. You may be said to have a property called spin. If you spin very slowly, let us say you face south at noon and rotate so that it takes twelve hours to return to facing south, others may not recognize you at midnight as being who you were at noon. It will take another full rotation, to the next noon, to bring you back, in their perception, to resemble the same person that you were the day before. The same nuances of shade under the brows, nose and chin and a slight squint to the eyes under the mid-day sun will return. In particle parlance you have what is known as 'spin 1/2'. Indeed you have, for all the matter of the universe is made up of spin 1/2 particles. Particles with the attributes spin 0, 1 and 2 contribute to the forces that bear upon particles of spin 1/2. Particles of spin 0 have the appearance of symmetry from all directions, as a sphere; spin 1 particles look like they face in one direction, like you and I; and particles of spin 2 look like push-me-pull-yous with two heads looking in opposite directions.

Conservation of Energy

Referring to the universe of all-there-is, and accepting the premise that 'all-there-is is all there is', we must conclude that no exceptions to this general condition are permissible. It follows therefore that any suggestion that there may be any more or any less of a totality is disallowed.

We now need to carefully define what it is to which we refer that does not vary in quantity. Insofar as we have attributed, and to some degree rationalized, that at the earliest moments of the universe all was energy, we may reasonably presume that it is energy that we refer to that exists in a constant quantity. We have thus arrived at a conclusion that there is a state of being whereby energy can neither be created nor destroyed. This status is accepted by science and referred to as the principle or law of the conservation of energy. Only relations between parts of a system change as determined by force, mass, velocity and time that are reconciled by the expression $F = mv/t$.

Forms

However, this idea of a steady state in the total quantity of energy does not reconcile with our experience whereby we witness expenditures of energy that appear to diminish residual reserves. Something else must be going on that interferes with our accounting system; and so it is. Energy has the property of transforming itself from one state or form to another, the effect of which suggests to our senses that quantities of energy change. What we are experiencing are changes in the relative quantities of various states or forms that energy assumes and reveals to us.

Released energy assumes the quality of heat, sound, light, electricity or chemical energy, whereas energy held in reserve in matter may assume the state of a plasma, gas, liquid or solid differentiated by temperature. The greater the density or mass of the storage form, the greater the quantity of energy held in reserve within a given volume. The distribution of energy within the electromagnetic spectrum (that includes visible light) is proportionate to wavelength whereby high frequency waves manifest greater energy levels than low frequency waves.

The more that energy is restrained from dissipating, the greater its potential to exert force!

Potential Energy

We opened this discussion by defining energy as the capacity to act forcefully. Insofar as such capacity is latent and held in reserve pending release it may be said to have an inherent potential for exerting force and performing work. Potential energy is a function of force suspended by a greater force pending release by an intermediary agency. The question whether such energy is potentially useful or inaccessible to man is incidental to the fact of energy being held in reserve. Theoretically all matter may be viewed as fuel. From a practical point of view, the utility value of a particular form of matter is dependent upon the accessibility of its potential energy.

Kinetic Energy

The second condition in which energy is exhibited is that endowed by virtue of motion. The quantity of kinetic energy in a given system is a product of its mass and velocity. The greater the mass and/or velocity, the greater its capacity to exert force or perform work. In common language, the term momentum refers to continuity of movement attributable to mass and velocity. We shall discuss this tendency, known as inertia, a little later.

Heat

The quantity of energy released in the form of heat is indistinguishable from the quality of heat since they are both attributable and proportionate to the intensity of motion demonstrated by atoms, molecules or other structural units of matter such that potential and kinetic energy cannot be identified as separate conditions while their sum total remains constant. Thus, heat and matter are concomitant. We discuss the effects of temperature change in the following Section on matter, and electromagnetism as a separate form of energy subsequently in Section 34 on force.

Equivalence

We recognize a correspondence between mass and energy whereby the greater a mass, the greater its capability to deliver potential, kinetic or any other form of energy. Our lives are inundated by evidence of energy conversion yet when we reflect upon the multiple forms of energy of which we are aware, we find that they are quantified in equivocal ways that lack correspondence. What we seem to need is a means by which to reconcile seemingly incongruent formats through use of a standard basis for measurement. Insofar as all matter is atomic energy in one form or another, it is reasonable to assume that a common unit of measurement may represent it. However, because the forms in which we encounter energy are diverse, solar, nuclear, electric and fossil fuels, etc, we tend to represent differing forms through the use of different measures of convenience.

Much as any form of energy may be theoretically converted to any other form, given the right conditions, so may such discrete forms of measurement be converted from one to another. One pound of uranium is equal in energy potential to 2.6 million pounds of coal, but in practice, unless it is beneficial to do so, we do not reduce them to a common expression, say watts! We shall speculate on the reduction of all energy to a single form in our discussion of theory in due course.

4   Matter: Predisposition Towards Behaviour

We have introduced the seemingly simple natures of time, space and energy only to discover that they are clouded by human interpretation. By contrast, the world of material things must surely be simpler. Substance is the real subject of predication. Matter is present or it is absent, is it not?

Common Understandings of Matter as Tangible Things

The everyday experience of the common man is that the presence or absence of matter may be confirmed by the senses, with some exceptions. We distinguish one material from another through perceptions of difference in the manner in which they stimulate our sensory systems of sight, touch, taste, smell and sound. These faculties afford us tangible evidence of matter as fact. Upon examination of that evidence we discover distinctions in variety and relations whereby some matter stands alone with the appearance of being pure and uniform in composition, which types we call elements, while other types combine to form complex unions that we call compounds.

Distinctions in Kind: Atomic and Molecular Structures

The idea that matter comprises minute building blocks was first advanced by Democritus in the fifth century B.C. His purely speculative thesis was that 'the only existing things are the atoms and empty space; all else is mere opinion'. Democritus laid out the general parameters of the atom (the Greek word for indivisible) as being size, shape, position and motion, which remain critical considerations in any scientific explanation of the atom. As is often the case in history, philosophical speculation about the existence of physical phenomena precedes confirmation by many years. Atoms were first confirmed scientifically by John Dalton in 1808 as a direct consequence of his discovery of the element, the atom being the smallest quantum of a substance that cannot be reduced into other substances, thereby qualifying as elementary. Atoms were found to be identical for a given element and the smallest quantity of an element that could take part in a chemical reaction. As with atoms, the concept of the element preceded its confirmation by science, having been articulated in 1661 by Robert Boyle.

Significant findings of Dalton included variations in the relative weights of atoms of different elements, from which he was able to compile the first comparative table of atomic weights. It is noteworthy to acknowledge that the only combination of the elements oxygen and hydrogen known at this time was water, which he found to comprise about eight parts by weight of oxygen and one part hydrogen. He deduced the relative atomic weights of oxygen and hydrogen to be in the order of 1:8 by incorrectly assuming a 1:1 match of atoms rather than, as we now know, two atoms of hydrogen and one atom of oxygen, which form a single volume of water and render a ratio of atomic weights of 1:16. Dalton had inadvertently denominated the ratio of the atomic numbers of hydrogen (1) and oxygen (8) as 1:8. The atomic numbers signify the number of protons in the nucleus of an element or, in neutral atoms, the number of electrons outside the nucleus, each element having a discrete number by which it may be identified.

A second and no less significant discovery by Dalton was the important role that numbers (that represent atomic weights) play in determining critical masses of elements necessary for chemical combination to occur. From this he devised his 'law of combination of multiple proportions' later confirmed by experiment, a law independently discovered by Joseph Louis Proust. A corollary that flows from the above findings is that compounds are formed from small numbers of atoms combining into what Dalton called 'compound atoms', which we know today as molecules.

The molecule, the diminutive form of the Greek and Latin words *Molos* and *Mole* for a large piece, mass or bulk, acquires molecular weight equal to the product of the weights of its constituent atoms. The chemical combination of compounds is determined, as with atoms, by the ratios of their molecular weights, but may result in the segregation and rejection of certain elements 'unwanted' in the new order. The notion of order is important in that chemical reactions are merely the rearrangement of a finite number of atoms. A molecule of common salt, for example, contains one atom of sodium and one atom of chlorine.

Nature's Basic Building Blocks

With the atom established as the smallest unit of a chemical element, and the molecule as the smallest unit of a chemical compound, we had descended to the bottom of the scale of material existences. Here at last were nature's basic building blocks. That nothing smaller was matter was a matter of fact. The search for the floor of chemistry appeared to bottom-out, fizzle and dissolve.

The continued search for the heart of matter is an unlikely story of particular familial relations that lead one from believing in simple notions to the brink of disbelief. We start the story with the revelation by Ernest Rutherford that the structure of an atom is a microcosmic solar system. Physics was accustomed to, and comfortable in searching for, universal truths in cosmic darkness and without apparent ends. It had been appearances of ends that had dampened chemistry's continued interest in the search. The prospect of new frontiers opening up for physics, at a time when the cloud of late nineteenth century opinion auguring an end to physics still hung low, was tantamount to an affirmation of reincarnation. Physics rushed in where chemistry feared to tread.

In order to continue the descent in scale of this line of investigation it is necessary to take leave of our senses. Sub-atomic physics demands that we extend faith beyond reasonable bounds in the absence of any abilities to confirm the veracity of what we are told. We enter an environment where strange ideas and higher mathematics fuse. Through the grace of physics the twentieth century has given birth to a family of sub-atomic particles. The word grace is used purposefully to suggest that a special kind of beauty attends veritable concepts. We continue to use the terms particle and matter while our common understandings of those terms become less and less identifiable with the findings of physics as the scales of the objects under investigation diminish.

Physics had become a quasi-religious inquisition for supernatural phenomena where each new conjecture portended a new divinity. Rutherford's discovery prompted the question 'what is going on in atomic circles?' Perhaps an answer can be disclosed by way of analogy.

The Big Apple

Imagine an apple with sweet juicy flesh enfolding a central core containing seeds and their protective shells in a kind of pod. There are small mass-less pockets that may be found in the apple. The seeds and their shells are the hopes of the apple for evolving and protecting future generations of apples. Hopefully, the future prospects are for more fine flavour and rosy colour. Compared with the parent apple the qualities of the future apple may range up in flavour to the top or down to the bottom of all possible tastes, be charmed or strange. The future colours are also uncertain; they may be green or red but a bruised apple can catch the blues. All depends upon the forces of nature. On the apple tree there is one force of nature that takes the form of large succulent caterpillars whose primary interests are devouring the flesh of the apple. The caterpillars are very positive about their mission but the flesh, understandably, is negative. The core may be swarming with latent destruction for even uncertainty has its degrees, absolute or partial. The components of the apple may be classified by science according to common qualities; thus, representative class distinctions may be by absence of size, clusters of seeds and shells, proximity of clusters to danger, or other considerations deemed useful. Thus, names signify classes determined by behaviour.

Now substitute atoms for apples, electrons for flesh, nucleus for core, protons for seeds, neutrons for shells, baryons for clusters, nucleons for pods, neutrinos for pockets, quarks for futures, positrons for caterpillars, mesons for dangers, hadrons for proximities, antiquarks for destruction, leptons for absence of size, muons for shortness of life expectation, spin for uncertainty, fermions for partial uncertainty and bosons and their cousins, gluons, for absolute uncertainty, while retaining the up, down, top, bottom, strangeness and charm of flavour, and the greens, reds and blues of colour – all while maintaining leave of your senses – and you will understand a little more about atoms and the relations of their parts.

Now substitute the concept energy for the concept matter, for this is what you would hope to gain from eating an apple which, by the time you have finished it, is, for all practical purposes, vacuous. What are we to understand about the up, down, strange, charmed, top and bottom qualities of the flavour of a vacuous apple? Or of the reds, greens or blues of its absent colours? Or of the integral or half spins of its uncertainty when its very existence is in question? Figuring this out is the joy of physics.

If one examines the structure of matter at the sub-atomic scale the most surprising revelation is that the properties normally associated with matter at scales that can be verified by the senses – visible forms and colours, hardness, etc. – no longer hold, and in their stead little void 'universes' are revealed in which the only detectables are centres and fields of energy. The apple has disappeared to be replaced by the electrical field. The particle and the charge are one and the same. Energy may be positive or it may be negative, but when they get together, that is where the action is.

All the qualities that we attribute to matter must be qualities endowed to matter by energy. Hardness, weight and mass are functions of how densely the atoms, primarily composed of protons and neutrons, are packed together and exerting their collective resistance to penetration due to the forces that bind them together and hold them apart, leading one to view matter as Einstein did in his special theory of relativity, as frozen energy.

The Role of Temperature in Instigating Change

The atoms and molecules of matter flux in a constant state of changing relations. Why, to what purpose or, more properly for purpose is too presumptive, from what causes and to what effects?

The kinetic theory of matter starts from the premise – may we call it an axiom? – that matter presents itself in the form of atoms and molecules. In accepting this idea the theory hypothesizes that the presence of heat is a factor integral to and manifest in the jostling of atoms and molecules.

The question that follows is whether heat is the cause or the effect of jostling. Whether it be one or the other, or both in a reciprocal relationship, the addition or subtraction of heat to or from a jostle speeds or slows the action.

Degrees of chemical activity are functions of degrees of temperature. The average velocities of atoms and molecules are directly proportional to the square root of the absolute temperature. More simply stated, the hotter it gets, the hotter they get, and the more excited they become as a result.

Temperature change is therefore the primary cause of changes in material states and relations at the particle scale. The effect of such continuing flux is to acquaint each particle with many others in the society of particles by bringing them into close proximity, through which process affinities are established. Unions are made between atoms and molecules much as they are in real life at life-sized scales, by getting to know of, and getting to know, prospective partners. So the purpose, if we can allow, or the effect of temperature is to facilitate the socialization of substances, the cycling and recycling. Nature is not so inefficient as we often like to believe, for all of nature's resources of yesterdays are present in today's *soup du jour*. We witness this process dramatically in volcanism where nothing is spared the destiny of being reconstituted by temperature and redistributed in the forms of gases, liquids and solids. Here we are hard pressed to distinguish between matter and energy. Matter and energy appear as interchangeable forms of the same phenomenon.

With the lowering of temperature the velocity of particles diminishes until they are disinclined to make unions but just congregate together closely, as if to conserve what energy they have left. To animate their conditions, they may be assumed to be in their wakeful states as gases, sleeping when they are in their liquid forms and hibernating in their solid states. Some substances skip the liquid state and transform directly between solid and gaseous states, a process known as sublimation. For example, sulphur, known to the Romans as brimstone or 'burning stone', and in the Middle Ages as 'phlogiston', was believed to be a manifestation of the principle of combustion. Much as the solid, liquid and gaseous states of matter are contingent upon temperature, so too is the fact of matter itself. Energy purports to relate to a state of movement in matter, while matter constitutes energy at rest. Matter becomes an expression of energy, and energy a function of matter. Matter and energy are reciprocal states, one being patent and the other latent at any point in time. They oscillate like an alternating current with changes in temperature, hot tending towards expressions of energy as particles of liquid or gas; while cold, the absence of heat, tends towards the expression of the same particles in the form of solid matter. The Big Bang Theory posits a time when all was energy and only after separation and cooling were the elements and matter formed. The ice ages on earth signal a diametric condition when more was solid and less was liquid than we observe today.

Distinctions in State: Solids, Liquids, Gases and Plasma

All true solid substances are crystalline; that is, they have natural geometric units of repetition. It is therefore useful to describe materials in terms of their crystalline forms to obtain a graphic picture of how each relates to its own kind and to dissimilar kinds and, as a consequence, how strongly they bind by virtue of their native formal compatibility. It is the performance of multiple parts working in combination that defines their architecture. There are just seven natural classes of crystal shape based upon their geometry. They are all axial but differ in the ratios of length and angles of intersection of their axes. All crystals have six faces (including tops and bottoms) with the exception of the hexagon, which has eight. The cube, of which common salt is an example, is symmetrical on all axes and all angles between axes are 90 degrees; the triclinic is characterized by three unequal axes intersecting at oblique angles. The remaining crystal forms fall between these two extremes. The arrangement of atoms in crystals of salt, with their 1:1 ratio of sodium and chlorine atoms, may be likened to a stack of cubes wherein each alternate cube has at its centre an atom of sodium and is adjacent to six cubes with an atom of chlorine at their centres, repeated to form crystalline structures of variable sizes resembling a single super-molecule.

Of singular significance to the shape of any crystal is its potential for extension in like form along any or all of its axes, thus enabling the natural or artificial deposition or build-up of a homogenic mass without scalar limitations and with no interstitial voids. The basaltic Giant's Causeway in Ireland and the Devil's Post Pile in California demonstrate that the strength of the system is inherent in the form rather than the scale. The combinatorial form assumes the formal characteristics of the single crystal. The architectural implications of such a system suggest a unitary expansion of built structures whereby a structure may appear to be functionally, materially, economically and aesthetically complete at any stage in its phased development.

Crystals facilitate the nesting of geometric forms in dense masses. Irregular volumes, when stacked, admit voids that result in less dense concentrations of matter. The forms of solids are generally stable, a fact determined by the density of their molecules held in equilibrium by the competing forces of attraction to, and repulsion from, each other within very narrow ranges of variation.

With increasing temperature the molecules become more energetic, their acceleration and boisterous movement extends their range from local play areas until, like over-exuberant children, they may wander beyond their abilities to return and become lost. The society of atoms is acquiring greater mobility. With each degree of temperature rise each atom requires more space, displacing its neighbours and rendering the material less dense. Emigration and plastic flow are progressive until, by degree, the liquid state is reached.

The density of a gas is in the order of 1,000th that of the same substance in its liquid state, from which it can be interpolated that for any given volume of liquid, let us say a cubic measure, where the density of the molecules is 1,000 cubic units, i.e. ten by ten by ten units, the density of that same substance within the same cubic measure in a gaseous form will be one by one by one unit at normal temperatures and pressures. The last state of affairs of substance that we note here, but which was the first through which the universe passed, was that in which temperatures were so high that they denied the formation of atoms, which state, called plasma, is the stuff of stars.

The effort of restructuring transformations from state to state requires transpositions of energy exponentially higher than for raising or lowering temperatures through a comparable number of degrees within a single state. Matter in the solid state 'tends' to hold its form and to resist rearrangement, matter in a liquid state 'prefers' to remain in a single unit but can be easily persuaded to disintegrate from homogeneous to heterogeneous forms, while matter in a gaseous state doesn't seem to 'care' about its appearances and may get ruffled without getting ruffled!

If time and space predate energy and matter, where did energy and matter come from and how? The simplest explanation is to suggest that they did not and cannot. They exist in perpetuity together. So now we have no need for divine creation, or a big bang to start the time clock; no need for the transformation of energy into matter as a precondition for the existence of matter since they are the same thing, simply presenting itself by degree in different states as two parts of hydrogen and one part of oxygen present themselves as ice, water and steam – the same H_2O expressed in differing concentrations of energy within any given volume.

Properties

Materials have an affinity for their own kind; that is, all else being equal they tend to congregate in 'families'. This tendency arises from the manner in which they are formed, the forms that they assume, their relative densities, surface characteristics, and their magnetic attraction or chemical composition. In nature sediments tend to accumulate in strata, subject to forces of interference, all according to the laws of gravity. There is a natural order. Thus, fine particles form clay, less fine particles sand, still less fine particles gravel beds, etc. Individual parts are worked by the forces of nature embodied in wind, water, temperature change and gravity, to effect changes in accordance with predictable rules. On occasion the strata are compressed, folded and geothermally heated to change their compositions, from which flow recombinations of materials with their own distinctive structural, mechanical, energetic and aesthetic properties.

Archimedes invented specific gravity. What does this mean? It means that he defined his observations concerning the relative densities of compared matter in such a way that his definition could be applied to gauge matter other than that he himself had taken into consideration. He established a standard by which material densities may be measured on the basis of comparison.

The thrust of science is relational investigation on the basis of comparison; that of art relational exposition. Neither is free of the other. The essential attributes of anything are, as determined by the comparative method, those characteristics that make an entity uniquely what it is rather than anything else, a theme that we shall continue.

Matter as Critical Masses of Packaged Energy

When a critical mass of energy aggregates in space to form a barrier to the intrusion of other energy, an entity is formed with its own identity and integrity. We call that entity matter while conceding that matter is nothing but energy that has passed into a particular category of form.

We are unable to discuss matter without alluding to its interrelationship with and proximity to space. Discussion of matter is dependent upon discussion of space; indeed, in our earlier discussion of space it was found necessary to raise the issue as to whether space was inclusive of volumes occupied by matter or external to them.

Matter does not exist except as an expression of energy. The existence of matter *per se* is confirmed by the existence of discrete, packaged reservoirs of energy. Energy does not exist except as it relates to masses of matter. The sum total of all positive and negative energy in the universe is zero. So there you have it, and there you don't!

Attending all experiences, and theses propounded to explain them, are apparent antipathies and antitheses; mirror images of existences that stand in opposition as a matter of principle, as good does to bad, being mutually exclusive and destructive. Matter has not escaped this fate although its counterpart, antimatter, is hiding, not in some rebellious state, but perhaps in distant galaxies of antiworlds and antipeople. The concept of antimatter is not simply a beautiful idea in support of symmetry in creation, but is supported by experiments that have revealed that for each of the family of sub-atomic particles there exists its nemesis the antiparticle, as positron is to electron, antiproton to proton, antineutron to neutron. But let us return to atoms, for they are the essence of matter and the quintessence of energy, quintessence being the fifth and highest essence in ancient philosophy, above fire, air, water and earth, that permeates all nature and is that of which the heavenly bodies were alleged to be composed.

Let us look at another example, the carbon atom. The carbon atom has six electrons buzzing around a central nucleus. Disregarding the complexity of their inner and outer orbits for the purpose of this illustration, the carbon atom may be likened to a sphere, the earth, in which a nucleus is at its centre, and six electrons are disposed upon its surface such that all are equidistant from their neighbours and from the nucleus, as represented approximately, for all is in motion, by the relations of Quito, Ecuador; Nunivak Island, Alaska; Tripoli, Libya; Bouvet Island in the South Atlantic; the Kermadec Islands in the South Pacific; and Singapore - to each other and to the centre of the earth. One might almost address each electron by such names. They are all held together and apart in a geodesic relationship by fields of energy. Each electron has its opposite number, as Singapore is to Quito, serving as counterweights to maintain momentum. While the energy represented by Quito is not insignificant, its mass or sphere of influence relative to those of its nearest but very distant neighbours, Nunivak, Tripoli, Bouvet and Kermadec, is such as to suggest that there is nothing much going on between. Any one, say Bouvet, is illustrative of the tiny and insular volume that it represents by comparison with that of the whole ball of wax – the earth. Such is the nature of atoms, overwhelmingly void and devoid of anything that one can call something. Atoms are slightly infected bubbles of space.

So we have as the primary palette of nature three entities: time, space and energy, matter being a compound expression of primaries, much as we have three primary colours red, yellow and blue, green being a compound expression of yellow and blue.

Physicists play with make-believe models of reality. Their toys are theories and the means of testing them. One such toy currently receiving attention is the String Theory that all particles are made of hyperdimensional space inferring that there may exist an inescapable connection whereby time and space, linked as space-time, are themselves linked to particles represented as matter but made of energy. The notion 'hyperdimensional' suggests that time, space, energy and matter are aspects of the same phenomenon viewed as different 'dimensions' along different axes, as limited visions of existence.

Part II

RELATIONS

5   Existence: Durable Being

What is the basis of existence? How should we define reality? If we embrace all that has had a detectable presence in the past (the 'ex' of existence as it were) and all that has being in the present (suggesting itself as the 'is' of existence), and we further qualify all such notions as having a presence irrespective of human consciousness, then we arrive at the concept of objectivity. If, however, we predicate existence upon human experience, then we admit two distinctly different forms of being: the sentient, and all else that may or may not be potentially perceptible.

The extent to which objectivity precludes subjectivity is a moot point, but if we combine the two indiscriminately, then we indulge a richer, more profuse understanding of existence. However, for the purposes of our discussions and in the interest of clarity we shall continue to distinguish between the two in order that we may define non-existence (including the future tense or 'tence' of existence) as that which is practically or theoretically inaccessible to confirmation or has not had being to date.

Anything with potential for being confirmed may be said to exist from an objective standpoint, as are hitherto unseen trees in the middle of forests, or moons circling distant planets. Anything that cannot be substantiated may exist as subjective realism, as is the case of love, the circumstantial existence of which can only be inferred as a manifestation of behaviour.

From the latter we must acknowledge that there are as many realms of subjective reality as there are subjects. There are also those domains of compromise between subjects that, while they may serve us well collectively, do not constitute either of the existences mentioned above. Common law, a contrivance of reason and emotion, belongs to this category. Laws do not exist by objective or subjective standards, but society is generally well served by believing that they do.

Existence as Human Experience

Subjective experience may be considered to arise from two sources: those that stimulate the senses and those that are conceptual in origin. Sensed stimuli are converted to concepts through the attribution of definitions. What determines that an original concept may function as a stimulus is the potential of the concept definition to illicit interest. The incidental function of stimuli is to activate concepts, while the utility of concepts is to stimulate action. The relation of stimuli to concepts are, thus, as causes are to effects. We call them by different names, as we do the chicken and the egg, because they have different forms and perform different functions, but they are complementary aspects of the same reciprocal begettal system, the means by which a state of being comes to be recognized.

The word 'human' may be understood as a noun identifying a class or a member of a particular class of animal life, or as an adjective describing a quality attributable to that particular species. Similarly, the term 'human being' may be understood as that same noun 'human' followed by the present participle of the verb 'to be', as the human 'being' active. But there is another expression of the term 'human being' in which the word 'human' is an adjective that qualifies the noun 'being', that is, something of the type human that exists. What is significant about this latter use of the term 'human being' is that it is felt important enough to stress human existence emphatically in a form that one would never elect to use to describe other things, as, for example, a 'monkey being', a 'tree being' or a 'rock being'.

What underlies this notion of emphatic existence is the unspoken understanding that human 'being', that is the existence of the human, stands in priority over all other existences in the mind of man since existence as an idea is inconceivable without consciousness, and consciousness is the special property of the human. Reality is totality, any detail of which, while it may seem real, is experienced in the limited context of virtual blindness to that total reality, from which the detail may be interpreted to be virtual itself.

Plantation

The tree in the forest that has never been sensed by man or beast may be said to have potential for or be a latent stimulus, while the tree that has been visited may be patently so. More broadly, existent 'things' stimulate action, reaction or thought. Thus, the tree stimulates water to be drawn up its trunk, evoking a physiological activity by which operation it may be understood to be a patent stimulant. The tree has no concept of water for it has no administrative centre of organization, but it would not be a concept itself without water. Water abuts roots that draw it up. The roots grow and feed upon more water. A prime cause of a stream of causes and effects takes effect.

Animation

The animal on the other hand is stimulated and has concepts, as of danger. Concept or perception for the animal is a stage of development beyond the simple cause and effect experienced by the tree whereby an impression of probable cause is connected to an impression of probable effect. Like the experience of the tree, the experience of the animal is subconscious, even when such experiences are products of learning. 'Theirs is not to question why, theirs is just to do and die', to paraphrase Alfred Tennyson from *The Charge of the Light Brigade.*

Being Human

As noted, the transformation of the stimulus image to the concept image relies upon the attribution of definitions. If a would-be concept, say a thump in the night, cannot be articulated in some coherent form it falls short of utility and simply remains a stimulus. Many so-called abstract works of art, regardless of motivation and artistry, cannot be accepted as meaningful by many observers unless the stimulus does something more than hang on the wall. They must either communicate the artist's intentions, as is the case with representational painting, or they must stimulate new ideas. Both are conceptual and therefore qualify as stimuli for further thought.

Concepts as abstract thought are independent of physical stimuli but cannot be detached from memories and hence models of the physical, which models are themselves conceptual. So abstractions are not ephemeral conjury but rather products of stimuli one or more generations removed from direct experience. Abstraction is, as its literal meaning signifies, addressing something out of context. With the formation of a concept or idea, a subject is formed, which subject may in turn stimulate further concepts or ideas.

Being Reasonable

From the foregoing we have defined a world of existence conditioned by human consciousness. Insofar as each human has unique powers of thought and judgement that are explicable within limited terms of reason and the emotions, we reason and emote that we have independent existences. To the extent that an individual lives in accordance with the dictates of his conscience, he is more clearly individual and independently existent. Logic and belief prompt the construction of scarecrows, the motives for which are underlain by, feed upon and feed fear. The conscience has no place for such constructs, or such fear, and scarecrows have no conscience.

All our conscious behaviour, and much unconscious behaviour, is governed by the will, even those actions that we find necessary and take under duress or extreme danger, being judged to be in the best interests of the individual under the circumstances. Our exercise of reason or deference to the affections issues from decisions grounded in values that are uniquely individual, that is to say, that our own assessments of truth are what make us what we are as existences, whether or not we are more fully defined.

Existentialism

Existentialism is a non-systematic attitude towards human existence. It differentiates between will and reason in human nature. In rejecting reason and metaphysics as the governing principles of human existence (by which we infer individuality), we agonize over what kind of milieu we exist in.

That agony resides in minds that are unable to conceive of everything within or anything without. Under such conditions the mind is the milieu. Isolation is total. Under such circumstances, when we are alone, and because we are alone, we invent companionship, icons and other fantasies in fearful desperation of being alone.

Sören Kierkegaard, the founder of existentialism, made a distinction between the existence of man and that of any other thing on the premise that man is free to make choices and that, in so doing, he is in some measure responsible for his own fate. To the extent that those choices cannot preclude him from his ultimate fate, death, but merely offer multiple paths to it, he suffers. Not only does he suffer on account of the inevitable, but because, through his choice, he may hasten the event.

Because we are burdened by choice, we assume responsibilities. Because we are burdened by responsibilities we are burdened by morality. Because we are burdened by morality we are further burdened by suffering. Through such torturous processes we prescribe and define our individual introspective theories of being.

Time

Time and existence are concomitant, not merely tethered together. Each is an essential attribute of the other. This raises the question as to whether they are not one and the same phenomenon.

Time, our time, is finite; death is definite. Such a limitation, considered or ignored, raises time to the status of essence, from which one must infer that to waste time is to add an additional measure of suffering. Death thus becomes the bottom line in accounting for lifetimes. As in commerce the bottom line becomes the driver and we, the people, the driven. In existence man is committed to a life sentence.

The life cycles of dreams, ideas, hopes, faith, belief, actions and the physical that they implicate are all real within their own frames of reference.

Space

All that exists has a nucleus, a centre of gravity, or a locus that defines their being in place. Each such place exists within a greater environment such that the centre of one existence can be related to that of another by coordinating lines, directions and distances. By such means all can be said to coexist.

Singular Being

The idea of unity prompts the idea of sharing, significantly the sharing or sparing of suffering from otherwise being alone. When one is young, energetic and eager to learn of new experience, it is easy to tire oneself with processes that exhaust physical and mental energy. But later, when time lies heavily upon us and we ruminate upon being alone, we feel the full force of isolation. Companionship, sharing and love, even the placebo of imagined love, serves to relieve the pain of solitude. In this knowledge, unity is all that we have; in pairs, in families, in brotherhood, in community, and in animated communion with all other existences, tangible or imagined. The closer the union, the less lonely the suffering.

Hyperbole

'To be or not to be, that is the question' is an improper question. 'Not to be' imports that whatever the 'it' of being is, it is not considerable nor to be considered. The act of posing the concept 'not to be' assumes a conceptual existence of non-existence. What the question appears to mean is: to maintain or to change identity or integrity, that is the question. But that is unsatisfactory for whatever 'it' is, the nature of its integrity will change, which change renders the question moot. We are left with the alternative question: to be conceptual or tangible, that is the question – any answer to which inevitably leads us alternately from one to the other.

Circumstantial Reality

We should consider for a moment what the concepts 'real' and 'reality' really mean in relation to existence. Real, as distinct from less real or unreal, draws its authority from the comparative context in which it is used. Understanding this is to understand that reality is not of itself the product of confirmation by the senses, or evidence to that effect, but rather a relative term indicating plausibility and fitness to circumstance. In this way we accord to a wristwatch the distinction of reality if it ticks and registers the passage of time, surreal if a draped impression painted by Salvador Dali to infer contradiction with its presumed physical function, and unreal if a photographic imprint.

Reality is a state of transitoriness in all things extant. All existences assume forms and all forms transform! Our place in the overall scheme is that of the only part of the whole establishment that is conscious that this is so.

6   Essence: Quality Intrinsic to Existence

Time, as we have observed, is that extrinsic quality that enables existence in general terms. However, existence is evident in inordinate number and resplendent variety. What can we understand from such observations that could lead us to explain such diversity? As we considered energy we came to view changes in form, as from plasma to gas to liquid to solid, as remaining unchanged with respect to functioning as reservoirs of energy while differing with respect to quantities of energy held in reserve within any given volume. We acknowledged these changes to be coincidental to critical changes in temperature. While we will no doubt agree that ice, water and steam are substantially (i.e. chemically) identical, we cannot claim that any one is the same as another. If change of temperature was the 'cause' of transformation, then change of density was the 'effect'. No change of chemical composition is technically possible that allows us to continue to call water, water. Water must be two parts hydrogen to one part oxygen. Here we should disregard that common language allows that we may qualify types of water with an adjective to distinguish waters that have been reduced by the addition of elements, as in the cases of soda water or heavy water.

Singularity

When we acknowledge that the temperature of liquid water ranges between one hundred degrees on the centigrade scale (or 180 degrees on the Farenheit scale) and does not substantially change when transformed to a vaporous or solid state, we are admitting that change in temperature is not as critical to water being water as change in chemical composition. From the foregoing we conclude that critical chemical composition is closer to being the essence of water than is critical temperature. We understand that what we have distilled from the above discussion is a critical factor that enables us to rediscover water at another time and place with a very high probability that it is what we think it is, having the same intrinsic properties as the water of our earlier experience. This is not a distinction arrived at by comparison between what water is and something else that is different, or of something else that is distinguishable by difference in location or time. Quality as it relates to essence refers only to the extent to which our sample conforms to our definition of essential properties or characteristics, compared with the extent to which our sample does not conform; as pure is to impure. Quality does not imply value. Essence is therefore an exclusive characteristic of a particular existence, gained by virtue of excluding all extraneous elements and excesses in quantities but those absolutely necessary for something to be what it is – exclusively. Here we have simply discovered in its chemistry the most important quality of water (and any other element or compound), its claim to a unique and absolute standard of existence. That special quality or essence of water is thus, its proprietary numeric relation of two parts hydrogen to one part oxygen. The essence as quality is therefore a function of special relations. Only after we have established such relations qualitatively may we use number to distinguish existence as quantity.

The Essence of Nature

The fourteenth century English philosopher William Ockham (of Occam) connected existence to essence by way of recognition that existence and essence are singular, inferring that they were both properties of the same oneness, by which reasoning he found them to be indistinguishable and therefore the same, independent of form.

Today we might argue that existence is the essence of nature. From such a position we can interpolate that nature (or any part thereof) is the product of its essential elements in existence; that essence is nature divided by such existence, and that existence is nature divided into its essential elements. More simply stated: **Nature** = **Essence** x **Existence**; **Essence** = **Nature/Existence**, and **Existence** = **Nature/Essence**.

What we call Nature (*in toto*) is the sum (quantity) of all things in existence as determined by their essence (quality). From the above we can conclude that anything that is purported to be unnatural is either misconstrued or non-existent.

Qualification

Quality is singular insofar as it refers to a specific condition, whereas relation necessarily requires two or more objects of reference. When the quality of 'highness' is ascribed to a mountain a relation is often implied whereby the word 'high' is intended to convey greater height than the land in the general vicinity. The term quality, however, can be misleading where used subjectively by attaching attributes that are not inherent in an object or its environment, as is the case of 'beauty', which is the property of a person making a qualitative, relational judgement. We have, thus, revealed a distinction between qualities that are verifiable by facts and those that are expressions of subjective, relative values.

Before one analyses something for the purpose of understanding its nature, one must observe it carefully and lovingly, for there is in its totality a quality that will be lost through the process of breaking it down into parts. That quality is its essence. We should not therefore look for the essence of an individual life in the heart or the brain. Essence in life is more analogous to soul.

While essence may be viewed as a quality, the converse does not necessarily hold. Quality, like essence, may be viewed as a natural disposition that directs existence to extend in certain respects rather than others, but unlike essence, quality is variable. Essence defines absolute existence, quality defines relative existence.

Of the dimensions of experience that we have visited we have come to recognize that the intrinsic quality or essence of time is duration, that of space is volume, that of energy power, and that of matter density.

7   State: Condition

The primary state to which all things subscribe and with which we have wrestled to date is the state of existence, which, while we accord it primacy because all things subscribe, is not all that is possible. Nothing is possible; that is to say, the absence of anything that can be called a thing is possible, as are the unformed and the unimagined. That is to say, the condition of non-existence exists whereby future states of existence and novelty are possible but do not currently exist, making any state of existence or non-existence conditional rather than absolute.

We have discussed state to the extent of showing in our earlier example that the structure that water assumes is conditional upon temperature. State therefore is a product of subjugation contingent upon environmental conditions. Something expresses itself as it does in deference to external forces that render it more or less disposed in some way rather than another were it independent of such constraints and influences. To suggest that something has a natural disposition to be itself is so only in the sense that it is by nature disabled from being anything else due to the conditions imposed upon it. Any natural disposition to be itself only holds after something has assumed its state and form whereby inertia inclines it to continue being and doing what it is and does.

In as much as state describes a natural or contrived condition that could change, as from sleep to wakefulness, or as a geopolitical area contained within a prescribed boundary, state may be viewed as a mode with less security in being than the object or objects to which it refers. State is like an adjective that falls short of meaning in the absence of a noun. State is not to be confused with form but can account for form as a qualifier of status. So the first thing that one might reasonably ask oneself upon making the acquaintance of anything new is: What are its conditions of subjugation; what encumbrances come along with this novelty?

Motion or rest, while often regarded as states, are here distinguished from states on the grounds that motion and rest are not properties of things but of their relations to other things. Thus, the state of motion does not alter the condition of a thing *per se* although the by-products of motion as, for example, frictional heat or centrifugal force may.

In conclusion, mode or condition of being is that nature imposed upon something by circumstances, as an extrinsic rather than intrinsic property. Thus, a political organization that has supreme civil authority and political power is so not simply because it embraces the embodiment of the ethical ideas and will of a community or tyrant, but because such status is permitted by external factors. In other words, the 'architecture' or state of a particular existence is subject to external relations.

8   Relation: The Association of Entities

Relations Generally

The preceding discussions of time, space, energy and matter are all about relations. These broad subjects are environmental in nature and serve as contextual coordinate systems for any discussion of specific relations. It is the purpose of these few pages to reveal the uniqueness of relations. The term relation is used here to describe connections between two or more 'things' considered together that would not be possible in the absence of mutual existence. We are not interested in 'things' *per se* in the abstract, but in their associations with other 'things'. It is not intended here to distinguish between relations on the basis of form, cause or effect, except as they may serve to illustrate what relation is or is not. The term relationship is used here interchangeably with relation to refer to the state of being related.

Relations Defined and Described

Relation is the default position with respect to unconsummated union, conjunction the being together as distinct from the being one. Relations may be viewed as compromises of unity. By this is meant that entities in the forms of energy and matter are separated in time and space.

Separation is the sole criterion that establishes the existence of a relationship. Relations are substitutes for correspondence or concurrence in time and space. Thus, we are told, at the instant of the Big Bang, time, space, energy and matter were unified. A fraction of a second later there was separation, at which moment there were distinctions between manifestations of energy and, somewhat later, entities of matter, all of which confirm what we have already presumed – that plurality and relation are born of the same event. In the abstract, relations confirm the communal existence of energy and matter. Relations are coexistent linkages. One cannot think about a relationship without implicating a plurality of existence. A relationship by definition implicates some thing or concept as it bears upon another, as referent to referend, which bearing is reciprocal but not of necessity the same for each. The nature of a relationship hinges upon relative mass, force and distance, to credit Newton.

Significance of Relations

The significance of a relationship lies in the areas of shared dominion between otherwise separate entities. We shall discuss the impact of two forces bearing upon a point in due course.

We must now ask: what is the gravity or significant bearing of the point upon the two forces? Are the events leading up to the connection, and/or those post-dating it, 'the' relationship, or is the relationship the point, meaning the coincidence of time and space, energy or matter?

We are given to understand that two similar particles of matter or energy cannot, according to Wolfgang Pauli's Exclusion Principle, share the same position and the same velocity, subject to Werner Heisenberg's Uncertainty Principle, both of which we shall explore a little further a little later. Taking these principles at face value for the moment, and for the purposes of this proposition, there remain the environments of time and space that lead us to ask: can two differentiated moments in time occupy the same space concurrently? Conversely, can any two points in space share the same moment in time? In reviewing all that has preceded this page we have established that none of the above events are permissible. Thus, the thesis that time, space, energy and matter could constitute a single unitary possibility must be rejected. Because such denial is finite, because unity is compromised so absolutely, separation is forced to exist. Either we have unity or we must logically admit to having a degree of disunity. The singular is exclusive of the plural. Because separation is forced to exist, relations of necessity arise. A relation of one thing is therefore proximate to something else, which answers the question above concerning whether relation is or is not the 'point'. If it 'was', there would be unity. Separation is a degree of lacking in correspondence that confirms the existence of a relationship.

The extent of interest or value each party to a relationship holds may not be the same. A dog relates to its tail and wags it. The tail relates to the dog and to some degree exerts inertial force upon, and therefore wags the dog. That the dog exerts more influence over the tail than the tail does over the dog speaks of the distinctive status of each in their relations.

So a relationship is not merely singular in location and time by straight alignment between two points, it is partitioned as between the interests of the parties. His interests in her cannot be perceived to be identical to her interests in him, which makes of what we call a single relationship, two; a two-way track or an alternating current on the same track. That they fall in love depends upon each making a connection, he with her and she with him. Those connections are relations. At issue now is whether a relationship is shared or individual? The sun at noon at Nunivak, when viewed at the same moment in time from Singapore and Quito, signifies a simultaneous rising and setting. What a concept, what a relationship, or is it two or more?

So we see that if a relationship is observed by others, that relationship becomes clouded by the perceptions of the third and other parties, whether they affect the observed relationship or not. So in one sense a relationship is the private province of a party to it; in a second sense it is the preserve of the second party to it; in a third sense it is the shared domain of the two parties, and in compounding numbers it becomes the interest of any parties that are in any way incidentally affected by it. So whose property is it? Are relations private or public? What is going on? What is going on is nobody's business; that is to say, relationships exist, period. If anyone is to get personal about them, that is their privilege or lot, by choice or circumstance. With respect to ownership, parties and relationships are co-owners of each other insofar as each has a potential bearing on the fate of the other.

Relations as Utilities

If one accepts the premise posited in our discussion of 'matter' that the effect of change is to facilitate the socialization of substances, and the message in the discussion on 'state' to the effect that state is conditional on the degree of subjugation by external parameters, then relations can be seen to represent statements of the status or static condition of any one thing vis-à-vis another at a particular point in time within the general context of change, as cross-sections in time or still frames in a movie. Relations do not change *per se*, they supplant each other in serial succession to form patterns. The now-familiar process of morphing, where a form is resolved into a different form in a gradual flowing sequence that gives the appearance of transformation, is just such a serialized event.

If you view a relationship from a position outside its immediate context, and in disregard for the general direction in which change is moving, the relationship has little meaning except as it may serve to catalogue a particular condition at a particular time. Relations are serialized events connected from past to present from which we infer futures. That inferred future carries a potential utility value of a relationship, from which we may assess the likely costs and benefits and intervene where it is in our interests and within our means to do so.

Mutual Impact and Scale of Relations Differ

Relation occurs where anything coexists with anything else. Such things include the physical: solids, liquids and gases, and the properties of the mental realm: ideas, beliefs and values. Where two dissimilar things impinge, the prospects of conflict and resolution are inevitable. It is only where man has potential for influencing the outcome that we worry about how we might influence the resolution to our individual advantage. The necessity for accommodation is obvious, as is the case where politics impinges upon art, for example; either the energy in each of the disparate forces can be unified and directed to a common cause, or they will be mutually disruptive. Typically they need to disengage and to apply their energies elsewhere. What relations evidence are bases for changes in form. This is the essential nature of relations. Where the entities involved appear not to impinge upon each other, as between distant bodies, their mutual influences may be too subtle to draw attention or to measure, as weak gravitational attraction, magnetic repulsion or kinetic energy, but the fact of multiple existences determines the existence of relations, which in turn manifests the existence of separate forces. If we acknowledge that multiple existences are responsible for multiple relations, we are not only accepting relations between entities but as by-products of these existences, relations between relations, and further relations between entities and relations.

What do we call a relationship between relationships? A system. And a relationship between systems? An organism. And a relationship between organisms? A life. And a relationship between lives? A family. And a relationship between families? A community. And relations between communities? We could continue, but what is apparent is that there are scales of interrelationship from the simple to the unimaginably complex.

Relations Between Relations

In order to suggest the nature of compound relations and their interrelationships we will refer back for a moment to preceding discussions to see in a few examples how evolution has moved the universe from the simplest of beginnings to an environment of ever-increasing complexity.

In Section 1 on 'time' we noted a conscious change of mind whereby, through observation and measurement, we became persuaded that time was not a static coordinate system but rather a dynamic of relational systems that took as its points of reference individual perceptions of fixity in each human mind. Time transpired to be personal time and behaved in a relation-oriented, relative and conditional manner. Similarly, when addressing 'space' we found that our geometric relational descriptions of the universe always put us, as individuals, at its centre, and all events were perceived to turn about and relate to us individually.

In considering energy, man has deduced that in the moments following the Big Bang at the beginning of our universe, the only distinguishable differences between particles were between photons, electrons, neutrinos; their antimatter; and a few protons and neutrons. The formation of these novelties coincided with the formation of new sets of relations. Following mutual annihilation, the number of surviving types of entities, distinguished from instances, was three: protons, neutrons and electrons. The number of types of relationship, also distinguished from instances, was six. The agency that brought them and therefore their mutual relations into being was a decrease in ambient temperatures brought about by the disbursement of concentrated heat. As we have seen, as temperatures further declined a new range of temperatures critical to the formation of new relationships facilitated the bonding of protons, neutrons and electrons to forge deuterium. As temperatures fell further other elementary particles formed, each of distinctive combinations of available elements, until temperatures lowered to critical ranges where variety in entities extended further to enable more complex forms and their relations to emerge.

Let us look at numbers for a moment. If there is a single entity there are no relations. Where there are two entities, let us call them A and B, the number of relationships between them is two, AB and BA, accounting for the relations of A to B and of B to A. Where three entities A, B and C exist there are six coupled relationships: AB, AC, BA, BC, CA and CB. Where there are four entities there are twelve such relationships, and so on, always in accordance with the equation $R = n(n-1)$, where 'R' is the number of resultant two-party relations and 'n' is the number of entities available to be related. A relationship may be likened to a linkage between two elementary particles, which linkage forms a new, compound entity. In tallying entities we must therefore also consider the number of relationships between relationships using the same formula, and then the number of relationships between combinations of relationships. So the fact of association, indeed the fact of multiple existences, generates additional entities as all parts relate to all other parts, to all conceivable groupings of parts and to the total community of entities. We can now see that with the addition of each new entity the number of paired relationships increases in geometric proportion.

Relationships form new entities each of which comprise an assemblage of simpler entities, as parts. Through the flux of expanding associations these become compounded and more complex. This raises the question as to whether all entities are relationships comprising smaller parts, or potential parts, including the ultimate entity, the singularity. If one has an apple one does not have a half apple or two halves until they are so defined. Science has not progressed to a point where we can say with authority that anything cannot be subdivided into smaller parts, nor can we place limits upon the extent to which parts can be combined. Such evolutionary (or are they revolutionary?) events suggest that sooner or later everything will be introduced to everything else, establishing affinities when and where conditions dictate. By this logic, extended infinitely into the future, not only is anything possible or even probable as permutation continues, but in time everything appears to be inevitable.

Now transfer this concept of compounding relations to the expanding numbers of people on earth and we get an insight not merely of unprecedented increases in numbers of entities, but of a compound effect of added relations far greater than the number of individual people involved. In human terms compound relations include couples, families, in-law's families, friendships, acquaintances, racial, ethnic and religious affiliations, economic classes, cities, provinces and nation states, individually and in combinations as classes. We begin to get a picture of exponential complexity that mere increases in numbers cannot suggest. These are just relations between entities of like kind, humans, and their aggregations.

Now stir in all other entities of energy and matter, animate and inanimate, and their compound relations in time and space. Consider that we are still in the early stages of discovery in experiencing these relations, as Euclid formally connected measurement to space and Newton distilled the relations of bodies to space. Consider further that the pace of discovery accelerates with each addition to the number of persons on earth and with each addition to our inventories of technological aids. Our inability to comprehend such complex matrices of relations blinds us to the inevitable impact they will have upon human mentality and the viability of future societies.

Relations Rationalized

In 1687 Isaac Newton published his treatise *Philosophiae Naturalis Principia Mathematica*. In *Principia* Newton advanced his universal theory of material relations in space and time, the gist of which was to contend that relationships between celestial bodies were governed by mutually attractive forces that increased in proportion to increased bodily mass and with decreased distance. Through this theory Newton was not only able to proffer his theory of gravity but to render a clear accounting for the apparent stability of the universe as a system of infinite parts positioned by the total effects of forces generated by each competitive mass with respect to its distance from all others. He described the architecture of the universe.

Relations are formal sets of mutual impacts. They do not require adjacency or close proximity but Newton's law of the square of the distance does seem to fit well even when applied to those relations where methods of measurement have not been developed, as is the case in assessing emotional relations.

The above notion of interrelations implies a proportionate bearing or influence of all things upon all other things. This idea has been expressed in accounting for weather patterns that owe their complexity and unpredictability to the compounded contribution of forces as marginal as the turbulence generated by the wings of a butterfly in Peking upon the weather in New York, which realization gave rise to the affectionate term 'the butterfly effect'.

All the relationships involving a single entity to a boundless environment are impossible to comprehend simultaneously. For this reason we cannot hope to gain a full insight into the state of our total environment at any point in time or to all possible consequences of changes that may occur. We can only hope to learn the principles underlying such interactions and thereby be alerted to probable future events from which knowledge we may take steps to mitigate the effects of future events deemed detrimental to our survival.

Popular Identification with Relativity

The theory of relativity has in its name the hallmark of every man's everyday experience. He understands in making judgements that he is making them in relation to his own personal experience. The word 'relativity' stimulates thoughts and questions concerning relations between things that may otherwise not be consciously associated, from which relational understanding, ideas and creative processes flow, whether or not the person has any acquaintance with Einstein's theories.

Insofar as we have no knowledge of 'nothings', we presume them not to exist. When we look at the very large we see an assemblage of many things, and their relatedness. When we look at the very small it is the same. In time we find ways of dividing even the smallest things into smaller, related things. We are unable to define the smallest units of existence or to say that they do not exist. What we have are relations, and relations to other relations, but no fixity. There is no absolute standard of motionless fixity except that to which each man attaches in his mind. Since such motionless fixity assumes a god-like lacking in confirmation; in the macrocosm of 'all-there-is'; all there are, are relations. Insofar as we have no knowledge of 'nothing', we presume it to exist.

Valuable Relations

Imagine that every 'thing' (represented as an idea by a word) is an independent star in the night sky. It is not difficult to recognize that under such circumstances there would be a separate and unique relationship between every star and every other star. If we are now to select a particular 'thing' (idea) with which we feel some affinity, and if we put ourselves in its place, as an animate representative of that 'thing', we will no doubt feel the influence of other things around us in differing degrees. Further, if we now recognize that all these relations are in a state of flux by which their mutual influences upon each other are changing, we must conclude that the relations of primary significance are those that bear upon the present, and those of secondary significance (to the extent that they can be assessed) are those that bear upon the immediate future. Since we each have a different relation to all other 'things', each will carry different priorities with respect to the ordering of significance to each one of us at any point in time.

If any 'thing', distinguished from every thing, was suspended in space and totally devoid of any relation, then, however magical it may appear if set in another context, it would be neither viable nor vital. Nothing exists *in vacuo!*

Geometry Rules

The logic of relations may be illustrated geometrically. The most cogent relation between any two points is represented by a straight line between them that forms a one-dimensional structure. Similarly, relations between any three points, if not in a straight line, are portrayed by the lines of a two-dimensional triangle. Relations between any four points are delineated by a three-dimensional tetrahedron unless they are in-line or fall within the same plane. Beyond four points relationships can no longer be represented geometrically in such a way as to infer that all points are equivalent and equidistant. With the addition of points, at least one point must necessarily be closer to a second than another, which event distinguishes relations by type, and suggests each relation to be singular and unique in identity.

While the essence of a relationship is a common connection, as a straight line is between two points, most relationships are more complex; that is, they have multiple connections in common, as would be the case in a human relationship wherein a separate common interest defines the essence of each aspect of a relationship. In this case, the common denominator with regard to mutual attraction may be sex, intellect, economics, etc., from which examples it can be observed that a common connection does not bear a point-to-point correspondence to interest in common, nor to like character; indeed oppositeness or stimulative potential frequently seems to be a common criterion.

9   Identity: Distinction in Existence

The beautiful simplicity of a single entity was compromised by its failure to maintain its integrity as an indivisible unit. Such failure, we have suggested, arose from the confluence of something and nothing. A dual existence transpired whereby the single component was not simply translated from the singular to the plural, but transformed into contradictions through the emergence of discrepancies.

From that point existence assumed a measure of complexity whereby multiple entities were differentiable on the basis of 'what', 'why', 'where', 'when' and 'how' they presented themselves.

Distinction

Occasioned by such circumstances, distinction between parts of the former whole became the only option. Once there is something existent, termination becomes problematic since, to the best of our knowledge, things don't simply transmute to become nothings. Notwithstanding conjury, they transform or cross over from the assumption of one form to the assumption of another or others. So once there was something, it follows that there must always be something or subdivisions of that something thereafter. The corollary would necessitate that, once there was nothing, and nothing but nothing, then, by the same logic, its rearrangement in form would dictate that it could only give rise to nothing, or no thing. Insofar as an entity exists, logic argues for it to continue to exist in some form or forms for all time.

Separation

With the dissolution of a singularity, differentiation is the sole surviving alternative. When there are two or more entities they are by definition discrete by virtue of their separation, whether or not the parts are intelligible. Each part may be said to have an id, or unconscious self, to distinguish it from any other. Thus, there is the entity with the id to which we could ascribe the symbol *A*, a second entity to which we could ascribe the symbol *B*, and so on, *A* being different from *B* or any other entity solely on the basis of separation in space and time. All other aspects of the entities could be abstractly and theoretically the same or different. We have made distinctions between separate things in space and time that have presented us with id-entity *A*, id-entity *B*, etc. Each is singular and unique. With separation established, plurality of ids and their relations coincide. Identity is the product of an act of discrimination between parts of a relationship.

Individuality

All identities must pass the test of physical or mental existence. A nothing has no identity since it cannot be discriminated on the basis of similarity or difference. While an identity will always have an essence as a property of existence, entities may share the same essence but each will maintain an individual identity. It is through the monopolization of a certain place at a certain time that an entity asserts its self-identification.

Identification

We have discussed identity in the broadest terms as individuality, that quality that holds a thing apart from all other things. While we cannot reasonably presume that the moon would lose or gain any degree of independence from the earth, or undergo any quantitative or qualitative change if man was not to look upon it, it is through the agency of man's sense of sight that he identifies distinctions between the moon and other bodies. Thus, identification is a function of recognition distinctly separate and different from identity; an extrinsic rather than intrinsic quality that is bestowed upon something by virtue of sensory perception.

Self-Identification

We can recognize what we call perception as being a human and animal faculty. While we may differ in our definitions of consciousness and the role that consciousness plays in perception, we have the clearest notions that we participate in processes of differentiation as human beings. The strongest evidence of this is our seeming compulsion to distinguish ourselves as individuals for the purpose of treating ourselves with special deference. This self-identification, this id of our own entities, operates at the subconscious level, making distinctions between positive and negative experience that serve as prompts to instigate our actions.

Self is that within us that maintains its presence with little or no change. It is that which is vitally precious to us insofar as it signals not merely who we are and distinguishes us from all other selves, but that we are, that we exist. In modern society we maintain facsimiles of identification like the personal name, the nominal relations to kin, the social security number, the passport, etc. in the interest of asserting that we are who we believe ourselves to be, individual selves with our own IDs.

Stridentification

Indeed, we often go to great lengths to promote our selfness through the way we dress, exhibit our taste, individualize our property, etc. We encourage others to read our particular or peculiar affinities for the purpose of id-entifying or differentiating ourselves from others in the best possible light, advertising our own perceptions of who we think we are, or vainly who we would like others to think we are.

Thus, the automobile bumper sticker 'My other car is a plane', as distinct from 'a yak', or no sticker at all, speaks of the self-identification of the bearer as being immodest (albeit sometimes with a touch of humour), or otherwise.

Auto-Identity

To what extent our subconscious id is simply mechanistic, as physical or chemical actions dictate, is an open question, but we do understand that plants discriminate in a similar way by utilizing or neglecting to use resources accessible to them. The symptoms evident in a plant's relative wellbeing or distress appear to suggest that it acknowledges its environmental relations on a subconscious, or perhaps we would prefer to say unconscious, pleasure or pain level. Plants are entities, they have identities and they make distinctions automatically.

Class Distinction

How we identify the intrinsic quality of difference in essence is quite a different question from how we distinguish between entities. What we often refer to as identification is not recognition on the basis of spatial or temporal separation for the purpose of confirming particular entities, or for making distinctions between entities on the basis of essence, but rather difference in kind to which an entity belongs, a more general process of classification involving differentiation by common and uncommon traits rather than by singular distinctions.

One way of articulating what something 'is', is to define what it is 'not'. This method cannot be viewed as satisfactory, however, since it cannot be applied to exhaust all alternatives.

Diversity in Unity

A thing, any thing, derives its identity from the unity of its parts and its relation to its environs. What we perceive that thing to be is that sensed, interpreted and presented to our consciousness. What we identify by name is a concept arising in our perception, not what it actually is, or is to itself. Each entity exists as an identifiable individual, whether we know of its existence and name it or not – and all in aggregate are one!

10   Prototype: The Original Product of Creation

In the idea 'entity' we confirm existence of a thing. In the idea 'identity' we confer individuality, as if to give it a soul. Like a soul, identity owes its origins to prior conditions that conduced its creation. Creation, as we shall discover, is the process of forces at work that give form to all things, which forms we hold in memory and name.

When a new form comes to our attention we check it against our inventory of known forms and make connections with those most similar. If appearances suggest that the new entity corresponds to one in memory, which memory is based upon an earlier model, we are not saying that the two are the same but rather similar in important respects. They resemble each other but retain discretely separate identities. The new entity is a look-alike copy of the older. The older entity is, if not original in an absolute sense, original in the experience of the observer and therefore forms the basis of comparison, as a prototype.

What is distinctive about a prototype is its singularity. Copies cannot be originals by definition. We distinguish prototype from succeeding like individuals, all of which have discrete identities, by its primacy in chronological sequence; however, a version succeeding a prototype may function as a later model and therefore be a prototype for its successors. A prototype is a design but does not necessarily require a 'designer'. Attribution by man to conscious man or God as a 'designer' is an exercise of judgement, verifiable or not as the case may be. Thus, 'design' is not necessarily an intentional process; it may simply be the product of natural processes, as arrangement.

Plato's reality comprised archetypal impressions that all like-experienced things closely resembled in patterns of relations. These models we might concede are in the mind as ideas and associations with which we compare copies, the first copies of which we often call prototypes also because they are the first demonstrations of such theoretical, abstract models.

The prototype is not identical to the later 'copy' but rather the conceptual pedigree that links the present to the past and so assures the observer a high degree of probability that the new will behave as the old. The distinction between close similarity and exact replication is important because, in the allowable deviation between the prototype and the copy resides the prospect and accommodation of special and conditional variations, and subtle changes as through evolution or development.

Broadly speaking, exemplary or original behaviour may be viewed as being prototypical in the sense that copying is intended or likely. Therein lies the value of the prototype, as a paradigm for change, from the Greek words *para* meaning 'akin to', and *digm* signifying pattern, model or example. As the copy closely resembles the prototype, so science endeavours to closely represent the truth.

11　Mass: Absolute Quantity

From experience we know that one cubic foot of concrete weighs, depending upon the mix, approximately 144 lbs . Why should we not be content to rely upon such linear measurements and weight? Do they not give us all that we need to know concerning quantity?

The answer to this question lies in the eccentric behaviour of what we call gravity, a force that undermines the notion of absolute weight. Weight, we discover, is changeable with respect to the distance of the site of measurement from the centre of a large body like the earth relative to which a weight is measured. How inconvenient, how confounding, how exasperating! The same cubic foot of concrete that weighs 144 lbs at sea level at the equator weighs 144.72 lbs at the North Pole! But we have to be careful here, for the one pound counterweight that we use as our standard of measurement will also change by the same subtle amount.

On such occasions that we need greater precision in defining quantities we therefore need another scale of measurement, one that negates the idiosyncrasies attributable to gravity, and another term to distinguish quantities that are measured on such a scale with the effects of gravity factored out. We use the term 'mass' to describe such quantities of matter.

Absolute Mass

We referred earlier to energy being potential or kinetic. From such understanding we can abstract the idea that the force of gravity is kinetic or active, rather than potential or static. If we now entertain the idea, if only in theory, that we can disengage all bodies of matter from the influence of gravity, we remove the tendency for each to exert forces upon all others and thereby to move relative to each other. In doing so we have transformed such forces from a kinetic state to a potential state to the effect of denying all bodies movement.

We may now make a distinction between a quality that bodies exhibit when they exert gravitational force upon each other which we call 'weight', and a second quality that shares all the same material characteristics, dimensions and quantities of the first but that does not exert such force. We call this second quality 'mass'. Weight and mass are reciprocal in the sense that weight is an expression of energy. As bodies accelerate towards each other, the mutual influence of their respective fields of gravity becomes stronger, demonstrated by an increase in weight. Mass, on the other hand, while independent of such increases, exhibits a quality called inertia, that we shall explore a little later, that subtly increases proportionate to velocity.

We can now see that it is useful to think of matter as bearing two separate qualities, weight and mass, that are made evident by variable degrees of motion. Each particle of matter bears both qualities, much as a partially filled drinking glass bears the quality of being partially empty. Neither can exist without the other. We can think of mass as the intrinsic quality of a body of matter that resists change in velocity.

Mass as Absolution

A dictionary will tell us that 'mass' is volume, weight, their combination – and a reverent chant! Could there possibly be a common denominator between the forces of nature and those formally invoked in paying homage to a reverence, and if so, what could it be that justifies the use of the same term 'mass'?

A few words concerning the Eucharist are in order. The idea underlying the Christian rite of thanksgiving is unreservedness. Much as we accept all things natural as given, we equate all blessings that we attribute to the supernatural with inevitability beyond our control. The connection and therefore the dual use of the term 'mass' is through analogy. To the extent that we fail to make distinctions between the two, they tend to become fused or confused. The relation between physical mass and Christian mass is as absolute is to absolution. Our lack of responsibility for such events is vindicated. Mass is a metaphor for absolute quantity with respect to authority.

Transformation

Mass does not exist. It is simply an expression used to describe the uniform behaviour of matter with respect to its transformation to physical energy. Mass in this sense is a convenient term applied to exhibit gravity and inertia as properties of matter related to its weight and volume. An increase in weight or volume, or in velocity according to Einstein's Special Theory of Relativity, contributes to an increase in mass, with corresponding increases in gravitational and inertial forces. Physical mass and energy are interchangeable.

The same is so with respect to mass as a reverent chant. It is simply an expression used to describe the uniform behaviour of supplicants with respect to the transformation of emotional energy – as behaviour *'en masse'*.

12   Equilibrium: Nature's Secret Force

Nature has created an unimaginably complex universality of disparate inventions, each with propensities for doing whatever it can do with whatever it has been endowed, subject to the accommodation of its neighbours and their endowments. The verb 'do' reverberates around the universe, and for everything that any thing does to any other, myriads of others do unto it. We have in place the appearance of a gigantic amorphism without a clearly defined direction, purpose or controlling influence, except as any part and all parts contribute to the whole.

We are digits in an unfathomable mathematical equation wherein every operation works to keep us moving, and every movement works to keep us stationary. The urge to go and the status quo cancel each other out, confirming the point of an equation – that it equates.

As we have acknowledged, the sum total of all positive and negative energy in the universe is zero, positive being that which acts in support of that which benefits from the action, and negative acting against. If there is any behaviour in the cosmos that we might feel inclined to attribute with reason to the will of a higher authority than our own, it has to be that expressed through the constancy of attempts to reconcile the equivocal, that called by the same name but differing in nature and function: the apposition of competitive forces, the reconciliation of which achieves a unity of parts that constitute a whole.

As the positive negates the negative, force to force, so equilibrium negates force as antiforce. The moral of this little tale is that we need to follow nature's example and seek a balance at every opportunity. The price of failing to do so inevitably is to invite hostile forces.

Every manifestation of nature appears to exist and perform within the confines of a limited range of possibilities. A balance point between the extremities of all such possibilities exhibits a state of quiet indifference or neutrality to the fury of opposing forces. Symmetry is not simply the condition that enables balance to exist as the centrepiece of action, it is the final arbiter of all conflicts. Equilibrium is nature's ultimate enforcement authority – her ace of trumps. To the extent that corrections are overplayed, events are played over until each storm abates according to the rules of the beautiful business of balance.

The extent to which nature tolerates deviation between static and dynamic relations may be considered to be the essential property of nature's liberal character.

13 Synthesis: Combination Effecting Change in Attributes

In the interest of clarity, let us make distinction between synthesis as the process by which nature combines the relatively simple into more complex assemblies and those processes that are the products of human will and action, intended or by default, to similar effect. The first class of combination occurs by force of circumstance and, if we are to continue to hold that everything will eventually be introduced to everything else, everything and anything may occur in time. The second class of combination is that which occurs due to human intervention in that otherwise natural process first described, not as an outflow of our being animals but by dint of our being cerebral and capable of conscious choice.

Here we shall limit our discussion to the natural order of combination, and call it synthesis, while deferring discussion of the acts and consequences of the latter to later discussions of reason and intuition as products of the intellect, and of will and decision as attributes of volition.

Natural Process

In describing the term 'relation' we used the concept of connection of otherwise separate things to show how a novel quality arises that was not in attendance when two or more things were not directly linked. Combinatorial Analysis, a branch of mathematics, has been developed to address possible permutations and combinations in terms of numbers. The principle was briefly illustrated as we explored relations between relations. Of greater significance is what could happen and what will likely happen, which the branch of mathematics we call Probability Theory treats.

The highest significance attaches to consequences of what actually does happen, the crux of which, as we have seen, lies in the actuality that sequence denies events just as surely as it enables them. These prospects cannot be revealed through analytical means.

The theoretical consequences of a big bang can only be conjectural, while the practical consequences can only be experienced. Mathematics cannot foretell with any degree of reliability where all the components of an explosion will be distributed at any particular point in time. Nor can synthesis by natural means be fully understood by looking at elements of a whole and projecting the most probable effects.

Even if one is to correctly forecast the future, say the weather, the correspondence between that which was forecast and that which later occurs is essentially coincidental; more probable than another outcome perhaps, but about as fickle as only weather can be - results compounded by the shortcomings of the analytic process itself.

For mankind, or so it would seem, synthesis must continue in the direction that it has in the past; that is to say, from the past to the present and on to the future. What measures of unity and variety are produced with respect to energy, matter and their relations will in all probability be those to which mankind will itself be reduced.

14 Integrity: The Unity of Parts

We will differentiate here between integrity as the quality of being complete and undivided, and the narrower meaning which, while it may embody the same essential idea of wholeness, refers to uprightness of personal character or spiritual rectitude, which we shall address later.

The Integrity of the Whole

One cannot examine the idea of integrity as a whole through differentiation and analysis of its parts since the concept of integrity is dependent upon the interaction of those parts performing in particular relations. Such interaction is absent from any set of relations other than that peculiar to the whole in question. To gain insight into the significance of unity one must examine the behaviour of the total assemblage of parts to discover what it can do that the unintegrated sum of its parts cannot. If it does not do anything more than a greater or lesser assemblage of parts, it cannot be said to have integrity except as an aggregation of more or fewer parts. We go back to the idea of essence to search for that which a critical mass necessarily has, or has capacity for, in order to become something that it is not in the absence of that critical mass.

Implications

Why, one may ask, should we be interested in integrity? Because it is integrity that governs performance, performance that governs action, action that governs achievement, achievement that governs advancement and advancement that nourishes self-interest. Once we understand that integrity advances our wellbeing we can look at the wondrous examples of nature with a view to modifying their performance in modest ways that will enhance our lives, increase the rewards for our labours, lend credence to our hopes and deliver our dreams.

Integrity is a state of being integrated, that is, of being composed and complete, undivided or whole. It is, for many, a utopian idea that seldom exhibits itself because, in life, all is accommodated by all else and in becoming so compromises perfection of fit and operation. Utopian integrity has, however, a more practical potentiality, that of giving direction. There is what might be called a 'rule of decrement' whereby the smaller the entity the greater the probability of achieving a state of integrity of its parts. An atom is, as far as we are able to tell, a stable and finite entity, the parts of which are interdependent to the extent that they cannot exist under normal conditions without being part of the greater whole. The molecule, similarly, is an integral exhibit by virtue of its reluctance to break down, or of its resolve to remain intact. Each is reduced to an amount that is sufficient to perform a novel function as a composite entity without excess. We need to distinguish between something merely working, and something working optimally, at which point it would achieve the highest degree of integrity possible. *Integer* is the Latin word for wholeness of number, from which we infer consummate performance.

Integrity of Quality

Consider a billiard ball. What is it about a billiard ball that makes it acceptable to be and do what it is and does and that gives it integrity? Minimally, it must have a specific weight, uniformity in material density, a specific diameter, surface characteristics that are invariable in sheen and texture, and a distinctive colour and number to differentiate it from others in a set. What the billiard ball has beyond its spherical shape is completeness without deficiency or excess.

Integrity of Quantity

This quality of completeness is physical integrity. Integrity as wholeness should not be confused with oneness. There are many billiard balls in a set. Each ball has integrity in form. Integrity in number of whole balls is a discrete and separate quality: three per set in English and French billiards (cannon or carom), sixteen in pocket billiards (pool) and twenty-two in snooker.

15   Order: Environmental Pattern

All has order. In the singular it is arrangement within. In the plural it is arrangements between. The reason that we are possessed by notions of order is that order is the principle by which we relate everything and every experience to all others in our minds. We are beset by an infinity of relations most of which have little direct bearing upon our daily lives within our comprehension.

Understanding is the key to making distinctions between what we call order and that which does not appear to fall into any such category of arrangement. Categories are mental files wherein we place like information. In the best of all worlds we would not need to classify and file for we would recognize and correlate all to all else automatically, discerning their correct relations, affinities and antipathies. Under such conditions we would exhibit a level of intelligence that would lead us to make judgements and decisions that were appropriate to each circumstance without inviting contradiction or conflict. This is the notion of heaven on earth that we entertain at extravagant moments but accept as inaccessible to humanity at the start of the third millennium A.D. We must therefore accept our destinies as matchmakers and filing clerks, devising tools to aid us in our deficient states, tools that often torment the tidiness of our mental aspirations.

Order implies border, and like all borders the lines that we draw to distinguish one concept from another are dependent upon imperfect judgements that remain open to question and revision. The fact of such imperfection negates the good intentions and dedication of the filing clerk. If he is intelligent, and if the issue is not one of life and death, he will reconcile himself to differences in opinion, understanding that neither he nor his complement has a firm footing in heaven.

Order as Arrangement

Order is an inevitable fact of nature misrepresented as an invention of man. It is a system of reference through which Alice, in projecting herself through the looking glass, may have reasonable expectations as she ventures into the future. What gives her reasonable expectations is repetition of experience in effects attributable to like causes. In this way she builds a library of experience. Such experience will likely but not necessarily correspond to that of others. The nuances of difference become the seeds of division. Systems are simply mental constructs that serve the purposes of assisting comprehension of related, interactive components. Systems are relational patterns. The wise will organize their differences into broad categories of agreement, the impoverished will fortify themselves against the mental precepts of the disagreeable and resolve to overcome they know not what. So there is recognizable order and that which is not, as past is to future. We embrace clear patterns and are confounded by incongruity.

Order as Command

'Order as command' demands the same measure of intelligent consideration as 'order as arrangement'; hence the same degree of understanding and the same name and spelling. Order as command invariably represents a demand for clarification of arrangement or pattern. Order as command is the automatic default mode of the mentally stressed, which attempts to bring within their purview that which naturally lies beyond. In this way order as command constitutes a signal event that the threshold of tolerance for disorder is at hand. Insofar as society is ordered, there is necessity for some degree of order as command. Children require the authority of their parents and teachers to guide their thoughts and actions. Groups working towards common goals need the authority of knowledge and sound judgement to direct their courses. If the structure of command is not open to improvement, disapproval will follow and tyranny will become the order of command.

The passage from order as arrangement or pattern to order as command is equivocal; that from order as command to ordination tenuous, and that from ordinance to ordnance treacherous.

Order as Confusion of Arrangement and Command

Imperial Rome, as evinced by Vitruvius, showed how military order commanded military arrangement from which the concept and term order was applied indiscriminately to both the processes and the products of military and civic architecture. We shall have more to say about style in due course. Suffice to say it is a short step from fusion to confusion.

Side Orders

One morning God was not observed sipping coffee in preparation for his day's work. No one has ever observed him sipping coffee so it was not an atypical morning. He was fully in command. The universe was in order. All was well. It happened that on that morning a man of some ambition was sipping coffee in preparation for his day's work. He was what we might call a unionist. In this regard he had much in common with God. The man was of the persuasion that if only the world of man could be raised to the level of understanding and respect that all was related to all else, all would be well. There have been others that have made significant contributions to understanding relations and their contributions to the whole, and he had set himself the formidable goal of organizing the disparity of parts into a comprehensible whole. It is one thing to work alone at the conceptual level towards feeling a unity with one's environment. We might call it a unity of mind and matter; one mind, all matter. We might think of such a condition as the ultimate state of order, tranquility if you will. The man, however, had a problem. The problem was of this nature. He recognized that he had satisfied himself of his place in the union, but was disturbed that others appeared to behave as if they had not. Not only had they not recognized the level that he had aspired to and attained, they did not recognize that they needed to do the same. His challenge thus presented itself to him in this form: how could he persuade them of the efficacy of his good thoughts in the interest of benefiting them through the enhancement of their relations to others?

The next morning he happened upon a good woman sipping coffee in preparation for her day's work. He asked her: 'What is your day's work?' to which she replied: 'I have been trying to understand why other people don't think and act as I do. My work is to bring their thinking and actions into accord with my own.' 'Excellent,' said the man. 'That is my most fervent desire too. What do you intend to do?' asked the man. 'To be alone,' she replied, 'for that is the universe where I can achieve the greatest success.'

Union is the order of one. Order is the union of many.

All is in Order

Everything is in order at all times. The order of the moment may not be that which one desires or can comprehend, which negative forms of order we 'dis' or prefix with disapproval and attach contrary meaning to. Disorder, by definition, assumes an order that is unwelcome, as weeds in a carefully tended garden are in contrast to plants of choice. Any state is as orderly as any other. The fact that order as we acknowledge it does not correspond to our expectations or preferences, or reaches beyond our means of comprehension, is a distinction in kind that denotes psychological boundary conditions within ourselves rather than in the definition of any particular forms of order, as 3, 1, 2 compares to 1, 2, 3.

If one is to argue the point as to whether all is in order or not, one would necessarily have to take the position of adopting a narrower definition of order, one that does not include any or all arrangements but rather only those arrangements that convey to us some semblance of regularity, progression or classifiable relation consistent with coherent systems of form or function. To take this position requires acceptance of the preceding proposition and a step across the psychological threshold to admit that order is in the mind and disorder that which confounds the mind and is denied; as 1, 2, 3 is to 3, 1, 2. In 3, 1, 2 and 1, 2, 3 we express sequence, to which we ascribe another name: order as arrangement. All is in order.

16   Entropy: The Tendency Towards Disorder

In discussing relations we looked at numbers to show how the compounding effects of additional entities increased relations exponentially. In discussing order as arrangement we implied that arrangement was a form of relation, from which by further inference we can deduce that multiple relations or patterns increase in geometric proportions. The discipline concerned with assessing the likely outcome to changes in relations is called Probability Theory, a subject that we shall address separately. Suffice to say, the greater the extent that events or changes in relations occur, the greater the chance that their outcomes will be unpredictable. To this phenomenal tendency for events to unfold in an unpredictable format we attach the term entropy.

Alluding to our discussion of arrangement and pattern, what science calls order is limited to that which is identifiable as pattern. All arrangements that do not subscribe to identifiable order-as-pattern must necessarily be represented by another term, 'disorder'. Order and disorder are the mentally secure and insecure regions of the greater relational context: arrangement.

Comprehension

Science reserves its use of the term 'order' to describe a narrower or subordinate concept, that of comprehensible order. In the general course of understanding we utilize ordination. Science subscribes to hierarchies or subordination, or what the God-fearing sometimes claim to be heresy or insubordination.

In the broadest sense of the term 'order', entropy does not exist. It is only in the limited sense, where science distinguishes between the comprehensible and the incomprehensible, that the idea of disorder and therefore of entropy is introduced to denote tendencies towards inexplicably complex forms and arrangements. The term entropy should therefore not be used to describe a scientifically sound fact of nature, but rather to describe a finding of science that acknowledges absence of discernible form, pattern, hierarchy or differentiation, or a condition wherein all is not recognizable as belonging to and behaving within the scientific realm of knowledge.

While entropy exists for those that insist that it does, for those that do not, it does not. Entropy is therefore not a physical condition of matter or energy but rather a state of continuous reorganization beyond the limitations of our capacity to formulate identifiable mental patterns.

In so doing we draw lines between those relations that we comprehend and those that we do not. Such lines are frontiers to our future understanding.

Rearrangement

Arrangement is all there is. It is general. Patterns are particular to specific arrangements. Patterns are all those particular arrangements with which we can identify. Entropy as a concept is science's way of telling us that we cannot comprehend all things at all times. That which escapes our immediate comprehension is obeying the law of entropy. If entropy is defined as tendency towards change and rearrangement, it does not follow that such tendency is towards disorder so much as change of order. If order is defined as comprehensible pattern, then all that is incomprehensible may be considered to be arrangement that embodies an order of complexity that has not yet been ordinated. From the scientist's point of view, entropy as an idea is secured by the second law of thermodynamics that suggests that any change in the universe is cause for rearrangement that translates to a loss of our prior understanding of correct relations. Since change is endemic we are beset by rearrangement. If we experience more change than we can manage we become disoriented, mentally rearranged or deranged. We have moved across the threshold from the realm of order as science would define it, to the realm of disorder, decay and chaos defined by those who can no longer mentally arrange images of physical relations.

Derangement or the Breakdown of Systems of Comprehension

In pursuing this direction of thought science poses a dilemma. Do such powerful tendencies towards change and rearrangement pose prospects for the eventual breakdown of all compound systems and their components? Of course, but the saving grace for mankind is that it may rely upon entropy and mental incapacity. The human mind, overflowing with incapacity and ignorance, is not bothered by anything that does not bother it! Out of sight, out of mind! What remains in sight and in mind we take great pains to rearrange within our minds into what science calls order. Thus, as long as we live, images of order will arise within our imaginations from a preponderance of disorder. Order exists and is redefined in its own image. Entropy notwithstanding, you may have your cake and eat it.

Entropy is governed by the same set of rules that governs existence generally. That includes the rule that the aggregate is greater than the sum of its parts. Where systems combine their measures of entropy, the gross amount of entropy increases. Continuing recombination ensures that this process continues. This is because the permutation of possible relations increases exponentially.

Entropy is the mathematical consequence of compound interest that leads us from order to disorder, or from one order to another, however one chooses to define it.

Intellectus

Stephen Hawking tells us that 'disorder increases in time because we measure time in the direction in which disorder increases'. A more consequential proposition would hold that disorder increases in time because we measure disorder in the direction in which time increases.

If we measure entropy by the degree of order in a system, we are predicating that definition on judgement that some measurable quantum of order is present and that it may be compared with larger and smaller measures of order on some scale. The real but unspoken definition of entropy can therefore be interpreted to refer to the degree of intelligibility of a system, which degree is necessarily contingent upon the degree of intelligence exhibited by an observer in assessing the system. Those who have a low tolerance for appearances of disorder seek to create order, while those who, in the opinion of others, appear to accept appearances of disorder as arrangements that have not been defined or named are less likely to be intimidated by the absence of apparent order. Thus, in entertaining entropy as an idea we are acknowledging thresholds of perception of order, and thresholds of tolerance for continuing the search for order. In so doing we distinguish between conservative and liberal mentality.

If we measure entropy by the degree of order (or disorder) in a system, then we must acknowledge that in doing so we have calibrated not only order itself, but also our own intelligence quotient! As we cross the boundary between the comprehensible and the incomprehensible, we cross between the orderly reserves of the potentially useful and the unfamiliar forms of the inaccessible and therefore useless, as one form of energy, say electricity, may convert to another, say heat, at the cost of loss of form, integrity and identity. Entropy takes care of its own!

The Threshold of Chaos

The salient characteristics of entropy are dynamic, nonlinear, aperiodic, finite and deterministic. Simple systems compound to breed complexity. While their driving forces may remain simple, enlarging combinations interject increasing uncertainty as to what they do or may do interactively. In the absence of comprehension of these qualities we are admitted through the whirly gates into chaos. Comprehensibility is to order as incomprehensibility is to chaos.

Aided by entropy it is always easier to make a mess than to make amends.

17   Priority: Order of Precedence

Priority suggests to us associations with time, sequence and values. Let us remember Hector and Achilles once more to represent that the effect of sequence or order of succession is to render subsequent events possible or not. Understanding this, we recognize opportunities to deny events that would be to our detriment and to invite those that would be beneficial. Thus, priority is an expression of personal or shared interests that we seek to interpose between present and future events to further our interests. Priorities infer alternatives, alternatives imply judgement, judgement heralds decisions, and decisions secure agreement or invite disagreement. From the foregoing it can be recognized that the establishment of priorities stands primary in the sequencing of events that lead to the unfolding of tomorrows.

Priority is, however, not an end in itself, nor a means to ends. It is signage to, and along the path to, futures. The 'signage to' addresses goals, the 'signage along the path to' signals ways and means. These distinctions we shall address in discussing intentions and decisions – further along the path.

18   Class: Distinction in Kind

Class is the expedient between the simplicity of singularity and the complexity of multiplicity. It is the device through which we make conceptual changes in scale in our minds between any and all, by which we create affinities between experiences based upon likeness of attributes (distinguished from affinities between polar opposites) as determined by our senses. Class is the product of aggregation whereby we project the individual into the group and reduce the many to the few on a comprehensible scale.

Limitations Between the Many and the One

In his Doctrine of Categories, Aristotle conceived not only of categories intended to signify predicates, but types of categories or classes of classes, from which our concepts of 'quality', 'quantity', 'relation', 'substance' and 'state' were later derived. Kant extended Aristotle's logic to make distinction between kinds of propositions and judgements that categories posed, which engendered further classification necessitated where expressions shared the same predicate and type but differed, for example, quantitatively. When 'the king is dead', 'the kings are dead' and 'all kings are dead' are compared, differences in logical form lead us to understand distinctions between the singular, particular and the general (i.e. all-inclusive), and by deduction the negative forms 'all other', 'some other' and 'no other' kings survived. What Kant was doing was attempting to move classification in the direction in which knowledge was inevitably to move, to form linkages between principles of an emergent natural science. Kant further asserted that the same categories could not be applied to areas of human experience that were not subject to experiment and observation: to wit the affections. While the intensification of classification continues in the sciences, philosophy has become more circumspect in its use, preferring to generalize distinctions, perhaps understanding that class is only as good as it is useful.

Transpositions Between Quantity and Quality

Social class is an obvious expression of human aggregation. Use of the term class to denote human status in society and thereby to infer relative quality is a natural extension of its use to differentiate between qualities *per se*. It is commonly a resort used to justify attitudes and actions that discriminate between peoples on the basis of difference for the purpose of gaining advantage.

The human is a species, genus, category or class of animal. We are all of the human kind. Insofar as sight is our primary sense, we can most easily make distinctions on the basis of appearance. Form, size and colour are the most obvious distinctions that we make, and deformity, degeneration and discoloration are the grounds we use to separate classes, as minorities from majorities. From appearances, which are quite varied, we cannot make distinctions between Jews, Christians, Muslims and others. It is only when we start to seek grounds for making narrower distinctions for the purpose to seeking advantage that we create classes distinguishable by behaviour and avowed commitments to particular beliefs. When we yield to our innate fears and predatory instincts and suppress our newfound civility we descend into lower classes of humanity. It is not difficult to recognize that we are dealing here with ordination, subordination and insubordination. Class is the broad brush with which we paint categories as groups, within which we create subcategories as sets, and subsets within sets.

The Reconciliation of Reason and Experience

Introduced to mathematics by Georg Cantor in the late nineteenth century against much objection, Set Theory, a bona fide theory of aggregates, has infused mathematics and logic ever since. We think of class, and at the smaller scale set, on the one hand as a contrived collection of distinguishables, and on the other hand as a singular grouping of sameness, but both concepts are captive subjects of our comprehension as to what constitutes difference and likeness. Underlying this reality is another, the bases of assumptions for comparison, by the individual and by groups of individuals, on matters of equivalence and the definition of difference.

In science a characteristic sequence of events occurs whereby observations are made, organized into classes by likeness in appearance, and ascribed laws to account for distinctions. Where such laws coincide, significance is attached because of their apparent unity of concept. As further observations are made they are tested for conformity in behaviour with expectations predicated upon the laws. If conformity is confirmed the laws are upheld, if not the laws are examined for possible changes in logic that would account for the new findings, or a new class and law are established to explain the newly found condition. In this way experience and logic wrestle in competition for validation. This may be viewed as the conceptual method of mainstream scientific investigation to which may be appended emphasis upon reason over experience by the so-called logical positivists, or vice versa by the behaviourists.

Relation as Facilitator or Inhibitor

We invent class as a means to understanding relations. At the most obvious and the most fundamental levels relation means physical juxtaposition. If we recall our thoughts about geometry and relations we remember that physical relations between three points in a straight line are distinctly different from those forming the apogees of a triangle. In a straight line A B C, B inhibits an unobstructed relation between A and C where in the triangular relation it does not.

At first blush the layman thinks of chemistry as that discipline concerned with changes in the state of matter. What we are less inclined to consider is that such changes only occur when there is correspondence in the mating of atoms capable of being united or separated. This is a function of numbers, not simply by virtue of their arithmetic 1:1 or one-to-more-than-one relations but also of their geometric relations, as A is to C in the above analogy. Odd Hassel showed how chemically identical molecules could have different conformations. Derek Barton went on to show that, not only did differences in the spatial structure or conformation of molecules lead to differences in their behaviour, but that organic molecules had a 'preference' for three-dimensional forms that determined their properties. Barton and Hassel shared the 1969 Nobel Prize for Chemistry for their contributions to this understanding.

So geometric relation may inhibit or facilitate other kinds of relation, like proximity and adjacency, to the extent that it can deny unimpeded interaction. The limitations of the A B C in-line configuration were overcome by the A B C triangular relation, and the A B C triangle's inhibitions were removed by the three-dimensional planes of the tetrahedron and other crystalline forms to which elementary matter subscribes.

A Class Act

This may have seemed a little esoteric to the average reader but geometry is governing our personal behaviour in much the same ways as it governs our parts. Take language for example. As we have previously demonstrated there are six possible sets of straight-line relations between any three entities. Let us substitute D, G and O for A, B and C and permutate them to see how they impact the mind: DGO, DOG, GDO, GOD, ODG and OGD. Geometry directs your thoughts, particularly if you are literate, and more particularly if you are imaginative. In this case DOG belongs to the same set as GOD. From a purely geometric standpoint they are identical constructs viewed from opposite directions as reflected images, particularly when viewed in the lower case:

doq
qod

There's class for you! It is the device through which we make conceptual changes in our minds.

19   Totality: All Existences

What do we know about everything, not just every thing (about which we can clearly not know all), but about everything as a totality? We have a name for such a concept that we call the universe, but then we set about placing limits on the extent of the universe by likening it to an inflating balloon. By doing so we set ourselves up for the obvious and problematic question: what is beyond the 'ballooniverse'?

Bubble

Three concepts have been presented as possible forms that the universe has assumed. The first is as extension infinitely in all directions; the second, Aristotle's singular, finite, spherical and eternal universe; and the third, the expanding universe theory. The first is attractive in explaining why an approximation of equilibrium is maintained in the relations of celestial bodies; the second and third are contingency explanations of the universe if the first is not as we have described it but has dimensional limits.

The two theories, the infinite and the finite, appear to be mutually exclusive, the one unlimited and the second limited in extent. But is this necessarily so? What set of conditions could reconcile the two theories without breaching the laws of nature as we know them today? We might characterize the unlimited sky as a steady-state theory.

Hubble

The expanding universe theory is the product of Hubble's confirmation that bodies in the cosmos appear to be increasingly distancing themselves from each other. Remembering Newton's discovery that it was impossible to make the distinction as to whether A moves around B or B around A, the same dilemma applies to our perception of whether distance between bodies is increasing or scale is decreasing, remembering that scale is a function of human mentality. If everything is diminishing in scale at the same rate about us we have no benchmark against which we can assess such occurrence. It would be as if we were speeding backwards down a polished tunnel, let us call it a telescope, in which all appeared familiar around us at all times, and only the light at the end of the tunnel, a distant star exhibiting a reddish hue, would tell us that we were getting further from it. By this explanation we could account for the shift towards redness in colour of distant bodies without the necessity to place limits on the size of the universe. We would simply be engaging in a mental phenomenon. We take Hubble's observations at face value and reason why what he tells us should be as he suggests. Anyone who seeks to verify Hubble's observations simply requires to condition himself to the same set of parameters that Hubble accepted.

General Conditions

In addressing 'identity' we viewed individuality as a distributive order of many, while in discussing 'class' we acknowledged categorical orders, each as a speciality within a larger context. To the extent that all individuals and all classes find unity in totality, membership in a collective order of one may be said to be the general condition. The universe is that of which there is but one all-inclusive whole.

With respect to relations, as we have seen, all parts or particles bear relation to all others, to all aggregations and to the whole, which relations define each part, each aggregation and the universe. We presume that all there is is all that exists, and that what exists does so on the basis of fact. We will examine the concept of truth later to see how we place limits upon its integrity, which places limits in turn upon existences, absolving the fundamental precepts of truth and existence from being absolute.

What is it that distinguishes the whole from the sum of its parts? What we observe is that a whole 'works'. It functions as a totality in a way that no lesser combination of parts would or could. In this alone it is unique as a kind. Thus, the universe may be viewed as a whole, or any part of it as a less complex whole according to a holistic hierarchy whereby each whole is dependent for its integrity upon its antecedents and sequels larger and smaller in scale.

A simple example, the observable effects of compounded energy, may serve to illustrate the attributes of wholes beyond those of their parts. The addition of energy serves to empower events that would otherwise not occur. The greater the force the greater the impact of that force. But it is also true that the greater the force the greater the reaction to that force, thus beginning the counter-swing of the pendulum of events from pole to polar opposite. The gravitational attraction of bodies towards bodies induces the slosh of oceans. Such is the power of the accumulation of parts we call 'moon', and its powers of attraction to which we attribute the tides.

Scales of Operation

For the origins of wholes we make assumptions, the most common of which is that in order for each whole to be composed of parts, each part must predate the whole in existence. Thus, our synthesizing minds lead us to believe that wholes are assemblies of pre-existing parts, as wheels, seats and spark plugs are to the car, while our analytical minds reason that the parts of a horse are not so. Are not parts as surely deconstructed fragments of prior wholes? No, one might insist, an atom of hydrogen is an atom of hydrogen, period one. But try to exhibit a single atom of hydrogen independent of all else. Try to divide it for better or worse. Subdivide its divisions. Indeed it is a whole, indeed it has parts. This is because hydrogen is a captive and integral part of that hierarchical system of existences, harnessed by the greater, harnessing the lesser, whether we recognize and classify it as doing so or not. How about that whole – the universe? Is the universe not an assemblage of moving parts? Indeed. The question bespeaks the answer. Wholes are always the sum of their parts, contributing something to a higher order of organization. Even the expanding universe, that distending mass of detritus, has its context, its greater environment. If wholes are defined by boundaries, then there are as surely extra-boundary conditions. By this logic the universe is a scalar part of a greater unknown whole, as a growing flea on a rhinoceros, and a rhinoceros on a continent, and so on ad infinitum, by which understanding we must concede that the so-called universe is so defined because we cannot conceive of a greater entity. Growth and dissolution are complementary cycles in an eternal reciprocating engine of changing relations.

Wholeness is Conditional

Examine, for example, the properties of wholeness that impress upon the parts. To be a car and to accelerate, a whole needs to overcome the inertia and friction given to it by its parts and their relations. Such needs govern the design of its engine and the gear ratios of its transmission system. In this case the whole is no more the sum of its parts than the parts are dictators of the whole. A discrete new relational property of the parts has come into being which gives the car, as a whole, autonomous if not automatic mobility.

The universe, we are told, is an almost closed system. Closed that is to all but radiant energy, which, by virtue of its velocity exceeding the speed of universal expansion, can enter or escape. Has anyone suggested that we should be able to see light emanating from outside the universe since radiation exceeds the velocity of the expansion of the universe by a factor of three? So, whether the universe as we define it is expanding or contracting, there is no way that we can tell the difference since the distant light at the end of the tunnel is always going to read red. The only difference, which we are unable to distinguish, is that in the case of an expanding universe the red light source is believed to be within the expanding universe, where in the case of the contracting universe the red light source could be external to the universe. We don't know if we are coming or going!

Wholeness is Conceptual

So at the macro level of existence no system is absolutely secure from external intervention or incorporation. Nor is it secure from within. We refer back to Plato's mental imagery of man's mental imagery that the philosopher is 'the spectator of all time and all existence'. But we go further, as he did without acknowledging so, and substitute the word 'speculator' for 'spectator'. Georg Hegel built his life's work upon his faith in reason, that is to say, all that which is the subject of belief is accessible to reason. He planted the dual connotation of spirit and mind in the word *'Geist'* whereby any who wished to emphasize the one was inextricably drawn to address the other as a necessary offset in order to gain a unification in understanding. The path from the ethereal firmament to affirmation passes through philosophy on its way to science.

The universe is singular. The idea of oneness is conceptual. In addressing totalities we can better understand the nature, essence and contribution of parts.

20   Variety: Diversity of Expression

The fact of existence, the essence of qualities intrinsic to existence, the states or conditions of being, and their relations, all contribute to separate identities. Such identities may be exclusive to individuals or generalized by class. In either event we have distinctions in kind. Such distinctions, differentiable by material, form, size, colour, texture, etc. and most particularly by what they each do in the context of time and space, makes of each a variant upon a theme. The general theme is existence, the subordinate themes are any number of classes, and the unique theme is the specific identity of each individual.

Compounding Variety

As we have discovered, it is in the nature of nature to compound complexity. 'Entropy takes care of its own!' As the universe expands and distributes its influential parts, the mix of matter continues unabated. Everything has a prospect of close proximity to every other thing and hence a potential for forming a new combination or variation upon existing themes. According to this hypothesis life was not merely possible, or even probable; sooner or later it was inevitable! The propagation of variety is the principle whereby nature enables its creations to adapt to changing circumstances without loss of diversity; indeed relativity compounds exponentially with compounding diversity. The engines of change generate opportunities that are heterogeneous and life-enabling, but inevitably, as by-products of that process, change delivers time-fused ends as surely as it presents beginnings.

Homogeneity

The cycle of change stirs all with all else into a possibility soup. From the prospect of man, possibilities are viewed as both positive and negative. His limited ability to comprehend and accept infinite variety and complexity prompts him to simplify complexity and urges him to purge the negative. In doing so he loses sight of generalities and focuses upon particulars. The particular consequence of so doing is to act in his short-term interests at the expense of the longer term, the outcome of which tends to limit rather than advance opportunities and, thus, becomes life-disabling.

Degeneration

We will not dwell upon the subject of variety here but take the subject up again when we examine life's processes and precarious tenancy with particular concern for mankind's responsibilities for and contributions to the counter-production of number and variety by design or default, otherwise known as corruption.

21   Number: The Character of Quantity

Much as it is reasonable to suppose that speech preceded writing, verbal counting likely predated its characterization in graphic form. Such developments logically arose out of a need to extend, indeed immortalize, human memory for the benefit of succeeding generations, and as evidence of transactions and other events that were held to be of communal and lasting value.

Number in the Abstract

Time permits existence, existence qualifies essence, essence defines identity, identity enables class and class aggregates number. Number therefore is contingent upon definition. Number is a descriptor of something else. In the absence of that something else, number is numb, devoid of sense, mute, dead or nothing. 365 = 0. Numbers are like energy, they may be positive or negative, but, in the totality of all that which is universal, they amount to nothing. Numbers may thus be shown to be deviations from the norm. The only absolute number is zero. What we call and use in the name of number are applied numbers, a distinction that can be seen between pure and applied science. Applied numbers are those endowed by us with utility value through affixing nouns: three turtledoves, two French hens and *a* partridge in *a* pear tree (where *a* = one). So long as you are not talking about anything in particular, feel free to substitute 0 for 365, and *a* for 1. Mathematicians do this all the time. For reasons of utility obscure to most of us, the telephone set equates *a* with 2.

The number 1 signifies existence, plurality in number signals duplication of existence. Zero by this definition is a symbol denoting the absence of number. 0 = non-existence. Number imports relation. Most notably the number 2 does not simply imply a relation between two entities, it explicitly dictates the existence of such relation. The number 1, it could be argued, bears no relation to anything else of necessity. To address that proposition we must understand what the number 1 is representing. In the abstract, as we have seen, 1 = 0. Zero *in vacuo* has no relation to anything.

But in the world of applied numeration, one is relative to something other than itself. It is tied to its noun. If tethered to something other than itself, it bears relation to that something else. Even one as it may be applied to the concept all-there-is, or the universe, has a relation to each of its parts. We should bear in mind that what we may apply the numeral 1 to may be construed to be the sum of many parts, or two halves, each of which may be described in the singular as one, or one of two, or one half. Thus, one is not only relative to the substantive to which it refers, but also to all other substantives. The number one therefore imports relation.

Numbers are symbols representing quantified entities or ideas. Being so, they can be reversed, or used to transpose ideas from one form to another. Thus, mathematics becomes the intermediary form through which conversion may be made, not merely by analogy, which is the province of words, but by analysis, which is a function of numbers. It was differing quantities that prompted the need for differing symbols to represent them, for practical purposes.

The Utility of Numbers

We have visited the clumsy, imprecise world of fingers, feet and cubits to show how, when resources available for manufacturing were confined to sticks and stone, a crude system of numeric approximation using cardinal or whole numbers was adequate to enable construction that still commands our awe and respect today. We have been to ancient Greece and enjoyed legendary stories of its geometers who transformed the simple sequencing or ordination of numbers to project them into multi-dimensional ideas. Pythagoras is remembered for his devotion to numbers, numbers he believed to be the language of nature. To an extraordinary extent he was able to demonstrate this.

In Imperial Rome, the absence of the numeral zero, and the limited understanding of divisibility, did not inhibit their domination of the civilized world. To divide and conquer was a function of practical power politics rather than theoretical mathematics. While the zero was used on occasion in Babylonia in the third century B.C., it appears only to have been used between other numerals and not as an end character since it merely denoted 'nothing'. The use of the zero as a digit to denote the tenth power of a number (as does the 0 in 20) had its origins in two worlds a world apart: that of the Maya Indians of the Yucatán about the time of Christ, and in the Hindu world several centuries later. The use of individual characters to represent discrete numbers greatly advanced computation. The grouping of Hindu characters in tens, rooted in the number of fingers in a pair of hands and used in many early systems, reached its modern form through Arabic translation.

In 1670 Gabriel Mouton re-ordinated numbers to give them an internal modular structure of sequence, a great advance from the Roman system of numbering, where progression from a certain date to the next would read from 'MDCCCCLXXXXVIIII' to 'MM' to describe the passage from the end of the second century A.D. to the beginning of the third. Mouton's advancement of logic decimated the improvisations and limitations of the ancient world to give us the decimal system as we know it today.

The imposition of logic to deliver a more powerful system of ordering quantities invited further examination. Experience had taught us about the infinite divisibility of time and space; surely, it was reasoned, this idea was applicable to the infinite divisibility of anything else that could be represented by numbers. Therefore, rationally, we should be able to so divide any numeral. With 'rational' numbers (positive and negative fractions) in hand, and with a neutral zero inserted between opposing numbers, the path was cleared for formidable calculation. Formidability demands relief and it came to calculation in the forms of the abacus, slide rule, mechanical, electromechanical and electronic calculators - and the computer.

We have witnessed an ever-increasing rise in the utility value of numbers. Human evolution is marked by an increasing facility with numbers. The processes of civilization carry a numeric accelerator component.

Numbers of Numbers

There are more numbers than entities by a large number, for they extend to infinity in two directions and, as if that is not enough, congregate in aggregations of all sizes, larger and smaller in quantity, which of course cannot be certain because they extend to infinity and define uncertainty, the most signal aspect of which is the absence of a largest and a smallest. We revisit a unitary scheme wherein we define and utilize the midsection of a system that extends to infinity in two directions, like time, space and the mid-range of the electromagnetic spectrum that we call visible light. All else is out of sight.

Doing a Number

To do is a verb. Verbs are working words. Doing numbers either denotes that somebody or something is working with numbers, or that numbers are working on somebody or something. In the case of calculation they are reciprocal. You get to the numbers and they get to you. It is not coincidental that, in an age where the populace bathes in virtual reality, mathematicians are immersed in 'imaginary', 'quarternion', 'hypercomplex' and 'transcendental' numbers.

Mathematicians design systems of numbering that can, in their imaginations, perform tasks that they have set themselves that more pedestrian numbering systems cannot address. So-called imaginary numbers produce negative numbers when multiplied by themselves. When measuring time using 'imaginary' numbers for example, distinctions between time and space disappear and space-time is confirmed to be a singular identity. By inference we are given to understand that 'imaginary' numbers are more useful and therefore more real than ordinary 'ordinal' numbers, and that imaginary time is therefore more real than so-called real-time since it is reconciled with space rather than excluded from it.

In order to survive in the everyday physical world, and to come to terms with appearances of paranormal mentality, we need to dispense with numbers and revert back to thinking about the things that they represent at the most fundamental levels. Numbers are not what we often think they are. They are merely patterns, patterns of 'nesses' (onenesses, twonesses, etc.). To say that ants have the quality of abundance makes no sense. Only numbers, representing quantity, indicate abundance. Those qualities that we equate with power, and therefore tendency, are functions of number not of the subjects to which the qualities refer.

A number is not a property of the thing or things that it represents; it is a property unto itself. That is to say, number belongs to numbers as a class. 'Ten green bottles sitting on a wall' refers to a group. Ten is the group commander. The bottles have no 'ten-ness'.

It is not mathematical prowess that you need to understand numbers, it is the imagination to perceive concepts and patterns of relations, and to attach significance to them. What we are doing is conceptually diminishing the size of integral parts. Integers no longer suffice to describe the integrity of parts. We are, therefore, anointing progressively smaller parts of parts with the quality of integrity while diminishing their quantities. The logical extension of this trend leads us to a world of infinitesimally diminished quality (of lesser parts rather than greater wholes) and to increases in quantities, a condition that we can readily equate with by transposing abstract numbers for real-time everyday practical experience with things.

Position

The repetitive use of line strokes to represent things by number dates back to about 3400 B.C. in Egypt, where ten strokes represented ten items. The proposition of prepositioning, i.e. representing quantities by distinctively differing characters, and ranking them in order of their increase in magnitude, is believed to have first been committed to symbolic language around 300 B.C. in India.

We shall examine proportion as it is manifest in aesthetics a little later.

Simplicity

In a world in which the complexity of numbers compounds and dumbfounds, simplification is an inestimably valuable means, and simplicity an eminently worthy end. To this end we have conceived of a system of ordering quantities with a great economy of character types. By combining the line stroke of the ancient world to denote a single existence, with the zero of the medieval world to denote non-existence, we can identify patterns of characters as belonging or not to particular patterns or sets. You will recognize that this is the binary language used by computers. In this scheme of numbering 0 = zero, 1 = one, and because we have exhausted the full range of digits in the binary inventory, we must now represent all other quantities by piecing or 'bitting' them together as distinctive combinations of 0 and l, or 'bits'. Thus, the quantity 2 is depicted as l0, 3 as ll, 4 as l00, 5 as l0l, etc.

One day someone will rediscover higher dimensions somewhat like the decimal system that comprises ten distinctive digital alternatives (rather than the two of binary systems), and devise a way of harnessing them to our economic advantage, and off we will go again compounding complexity until we are again exasperated and forced to seek a higher (or is it lower) order of simplicity. This is the idea underlying the notion of quantum computation that utilizes 'quantum bits' or 'qubits' and quasi-particles called 'anyons' that only exist in a two-dimensional world. Edwin Abbott (who we shall meet later in our discussion of sight) would readily recognize that we have returned to 'Flatland', his fantasy realm of two dimensions.

Are we degenerating towards single representations to denote values that vary with each use and user, coded like poetry and perception to monotonous bleats having peculiar meanings to peculiar people that are peculiar to special places at special times, for which sounds like 'fine', 'oink' and 'moo' are already reserved?

We promise not to bring up the subjects of 'qubits' and 'anyons' any more. Numbers are like ghostly companions that accompany all things at all times.

The Language of Unification

Mathematics has assumed the function of uniting otherwise irreconcilable ideas through their reduction to numbers. Through digital reduction patterns of relations between the abstract and the physical can be compared in fields as disparate as music, the visual arts, architecture, science, indeed almost any subjects that would otherwise bear little or no basis for comparison.

By such means our inventories of relations expand exponentially with corresponding opportunities to attach value and meaning to them. Numbers thus assume a utility function second only to spoken and written language as the medium of choice by which we aspire to communicate and share our life experiences.

Do unto numbers as you would have them do unto you!

22   Singularity: The Quality of Oneness

If, as we have just discussed, we find merit in the reduction of complexity to be a cause of peace of mind, how much greater would the potential for peace be were complexity to be rendered to a single unified state wherein subject and object would be the same.

Whether one looks from the scale of man upward to planets, stars, galaxies and the universe, or downward to atoms, electrons, neutrinos, gluons and mesons, all are being what they are and doing what they do within strict parameters circumscribed by their natural environments. In this sense there is unity in their being and doing together. They are all subjects of a greater oneness.

Unity

If we should interest ourselves in such unity we have precedence for such concepts in our predication of condensed forms that we euphemistically call black holes or singularities, and in the formation of that single concentrate that we hypothesize must have existed prior to the Big Bang and universal expansion. At some place in time, all energy and matter are presumed to have been united as a single entity. If we can plausibly theorize the nature of such a unity then we will have distilled the grand unifying principle of the universe, that is, what makes it 'uni'.

A theoretical reconstruction of unity requires that we turn the clock back and examine the emergence of the universe. The search for a grand unifying theory of the universe is underway and has been characterized as necessitating an understanding of the relations between the four fundamental forces attendant in nature: gravity, the weak nuclear force, electromagnetism and the strong nuclear force.

Disposition

We contend that we have an understanding of the sequence in which each came into being following the Big Bang. The key to an understanding of the construct unity resides in our reconstruction in the reverse sequence. Let us reduce the problem to a process familiar to many. As you heat popcorn on the stove the first noticeable event that you witness is defiance of gravity; the popping and jumping. Your action has led to momentary and repeated violations of the force of gravity, in theory and in practice. Now turn the heat up a little. What is the next revelation, a sensation of burning perhaps? You have broken the bonds of the weak nuclear force that binds atom to atom. You are reorganizing nature's building blocks according to your own plan. Physics and chemistry become indistinguishable, you have unified them. Now turn the heat up further. The pan melts. If you had been able to control the experiment and drive the heat up high enough what would have happened next? You would have spun the electrons off their orbits around the nuclei of whatever was left to 'cook', thus, defying the laws of electromagnetism. When you get to the last step in the integration (or disintegration) of the four forces you participate to the fullest degree possible. More heat, much more heat, will disengage the quarks comprising the protons and neutrons of atomic nuclei and the strong nuclear force will have been overcome. All forces will have been resolved into a single entity. The attractiveness of your enterprise will be irresistible. Needless to say, it is beyond our means to convert all this theory into practice, as indeed it is beyond our wishes. But in principle the processes of unification, distinguished from those of disintegration, point towards a simpler, more easily comprehensible, and therefore a more stable, existence.

Observations regarding the interaction and interrelation of matter compound to suggest a profound and revolutionary idea: the interdependence of everything upon everything else. It is the 'working towards' that we draw benefits from in life, not the 'ending of'. Goals are our motivations, the means used to progress towards them are our motive powers.

23   Duality: Opposed Conditions

One of the earliest apprehensions of emerging man was distinction of difference. Sensation signals it, perception identifies it, judgement relies upon it and decision causes actions that are intended to exploit it. Almost all perceptual experience seems to confirm the existence of a broad range of possible variations that lie between polar extremities.

In ancient Chinese philosophy this phenomenon was accounted for by the belief that all things accommodated elements of yin, the feminine, negative, passive and weak principles, and yang, the masculine, positive, active and strong principles.

Under the Yin-Yang doctrine all universal experience was explicable as the product of interaction between these two forces. The power of this philosophy lay in the fact that it accommodated the coexistence of what would otherwise appear to be mutually exclusive ideas, enabling contradiction and harmony, and multiplicity and unity to coexist, and in the fact that the common man and the intellectual could readily identify with such an explanation as plausibly accounting for his personal experience. From this understanding one could entertain the idea that change was simply a natural, often cyclical process of the realignment of forces, always governed by the same principles.

Threads of Evidence

What is one to conclude about this ever-present duality that insists upon foisting itself upon our attentions at all times and in so many ways? All that presents itself in the form of a duality does so without dependence upon any systematic, logical or scientific rationale. All such experiences may be characterized as being in linear relation between extremities that are in contradiction. Such impressions are not diminished by being reduced to systematic, logical or scientific rationale but rather, to the contrary, are clarified and illuminated. It is as though the first person singular is always on-line. One cannot get away from one's self-involvement in juxtaposition between extremities of experience. We are as individual beads, threaded on multiple lines graphically represented by dandelion seedpods, whereby our positions on all subjects are always to be found somewhere between extremities of what is possible. We experience our sense of relation to all such phenomena through referral to ephemeral points denoting the ends of ephemeral lines. The notions up and down, left and right and east and west are not simply physical, but mental also. In our sensations of place within society, in our religious and political positions, and in our commitments between conflicting loyalties, we covet fixity.

We may impulsively consider that we are always at one extremity of such dualities as sex (being either male or female) or of mortality (being alive rather than dead), but is this really so? Do one male and one female not equally compose us and do we not, to a startling degree, give issue equally to baby males and females regardless of our individual sex? Do we not divest ourselves of, and invest in, atoms at a rate that nearly replaces all our bodily parts in one year? How can we be sure that ten years ago we were who we are today? If not, what can be said in support of our continuity through time? If the person of yesteryear has dispersed, what can we understand about the state of mortality?

Relativity

We always seem to hold a position relative to everything else, whether we are conscious of it or not. Duality defines the limits of what is possible in relative terms. Relativity has governed our thoughts and actions for so much longer than we have recognized her embrace.

Even in the total absence of life, distinctions exist between senseless and insensible heterogeneous phenomena based upon irreconcilable principles. Time gives utterance to events in such a way that substantives are dependent for their existence upon compensating counter-forces that offset light against dark, hot against cold, good against evil, matter against antimatter, etc., as seeming necessities of existence. The interaction between these contradictory forces is satisfactorily accounted for as nature's means by which approximate equilibrium can be maintained and continuity assured. Diametric contraposition between opposing forces is a general condition. A cursory reference to any one of scores of thesauri will disclose an equal representation of corresponding and contradictory ideas masquerading as synonyms and antonyms.

It is as if each experience has its counterpart and that truth is not correctness *per se* but rather confirmation of the extent to which phenomena correspond to one's perceptions conditioned by experience, expectations or preferences, beliefs and the imagination – there being a natural and therefore human tendency towards resolution.

24 Plurality: The State of Multiplicity

Insofar as we have recognized duality in opposing conditions, we have acknowledged multiplicity of expression. By classification we have differentiated between multiple kinds of expression on the basis of distinction, and by measurement and enumeration we have quantified such qualities.

Origins

Monism and pluralism are theories of substance that make distinctions between the one and the many. While philosophers split hairs with regard to making fine distinctions in kind or class, William Ockham contended that 'multiplicity ought not to be posited without necessity'. We shall therefore confine this discussion to the concept of plurality rather than to infinite variety.

Causes

Natural causes of plurality are due to changes in the properties and relationships of whole numbers of things. These changes occur without regard for the presence or absence of consciousness.

Rabbits don't multiply. Procreating rabbits divide the genetic materials of their individual assets, subtract from their total stock holdings, and, through addition, cause an effect that we improperly attribute to multiplication. Rabbits are automatic, arithmetic mechanisms that configure in real numbers involving division, subtraction and addition to the effect of presenting a false impression of multiplication to man due to their advancement of pluralism. This is characteristic of life's processes and, not coincidentally, of the process of entropy.

Objectivity

Conceptually plurality may be said to exist in two general categories. First, matter that exists in its own right as a physical phenomenon satisfies its own claim to existence without human intervention: 'a rose is a rose is a rose'. That there are varieties of rose in no way diminishes the qualification of any or all roses to their claims to existence or to their states of plurality. Such things exist regardless of whether man discovers and verifies this to be so or not.

Subjectivity

The second general category to which plurality conforms comprises all states of multiplicity that are constructs of mentality. Such existences, whether in the singular or plural states, are conditional upon assumptions that may or may not be demonstrated to be correct. They belong to the realm of ideas that, dependent upon their credibility, find a place in belief systems.

By the first standard, we might consider all such things to occupy space, be potentially verifiable and thereby have their existence and plurality confirmed on the basis of what we call 'fact'. In the second case, existence and plurality cannot be claimed to occupy space and, thus, may or may not be capable of being represented as being factual. As to whether it is necessary for knowledge to be demonstrably factual to qualify as being true we shall leave to a later discussion.

We have, thus, arrived at a state of mental duplicity whereby we assert claims to existence and plurality of things whether or not they exist on the basis of fact. Thus, ideas and the objects of such ideas exist independently, conform to different standards and subscribe to different definitions in order to qualify for existence, a confusion that we shall revisit a little later.

25   Coincidence: Synchronicity, Contiguity and Accordance

Having established the nature of existence in the plural, we must face the possibility that two or more existences may occur simultaneously in time, space or in the minds of men. When we look for a defining idea that establishes synchronicity or coincidence in time we are led back to the admission that time is personal and that what we are trying to establish as a matter of fact is so only in an individual mind. In attempting to describe simultaneity we are embarking upon an undertaking that is technically impossible. Insofar as each of us is his own time machine, one cannot be coexistent with another in parallel time.

The same holds with respect to place. Contiguity, in the sense of being in the same place at the same time, is denied us. If so, how can we entertain such an idea as coincidence?

It is only by agreement that we can enjoy the notion of coincidence, by agreeing to pretend that we share the same time and the same space concurrently. That is coincidence by proxy, the only form it can take. If we accept that coincidence is a product of accordance, we must also accept that the constituents of coincidence, synchronicity and contiguity are equally inaccessible. They are a little like the notions me, myself and I, all referring to the same subject, in this case coincidence, but differentiated by syntax or context, by which distinctions we know what are being referred to, temporal, spatial or subjective experience as the case may be.

Coincidence, as commonly understood but less commonly acknowledged, is agreement that entities may share the same experience in any or all of the above respects, even when we know that they do not. What a coincidence! The incidence of coincident events that occur at any moment in time are innumerable. What is significant for us about coincidence is our growing awareness of it. Our perceptions of coincidence compound proportionately to our evolving capacities to detect them as determined by neurological, social and technological developments.

26   Change: The Equilibrium Equation

Of the two great universal tendencies – to continue doing exemplified by inertia, and to modify doing exemplified by change – the most cogent characteristic of the sensible universe is its propensity for change, while unremitting change is a manifestation of continuity.

The term 'change' as used here refers to a transformational event where distinctions in form, quality or state arise between prior and subsequent conditions. No attempt is made here to differentiate between change as evolution in the Darwinian sense, which we address later; evolution made manifest by flow, as of a stream of successive compounding causes and effects, or novelty with respect to new acquaintance that we will address as a function of knowledge.

Exchange, referring to those changes that are limited to reciprocal transfer, is the subject of our next topic.

Biased Change

The Greek philosopher Heraclitus, the earliest proponent of paradoxical logic on record, held that there is a discoverable system of correction whereby change is directional and directions are reversible. There is, he held, a built-in tendency towards the equation of forces for change with their polar opposite forces.

Parmenides showed how the then-current Ionian view that everything was made of a single substance was flawed, for if so the phenomenon of change could not be explained, from which finding the notion of plurality of substance emerged.

Sensible Change

Anaxagoras claimed infinite variety in substances. He argued that all things were made of all things, each varying only in its mix proportions, and that all was always accessible through division into smaller portions. Underlying his understanding of the nature of physical reality was his belief that what one sees is, and what is, is verifiable by the senses. His thesis allowed for no ambiguities, no doubt and no deceit. The idea that perceptions were not so fundamentally informative but relied upon impressions of form, quantity and arrangement of matter had to await the dual events of Anaxagoras's death and Plato's birth, both of which occurred in 428 B.C. With the death of Anaxagoras the idea that change was brought about 'by' nous or reason (derived from *noos* or conscious mind) was supplanted by the idea that change occurred for reason or reasons. The doors of perception were imperceptibly cracked open to invite enquiry as to what such reasons could be, and how such realities could occur.

Relative Change

Plato, in logically extending the theories of Heraclitus, maintained that everything was in a state of change. Such logic undermined the idea that what you see is what you have. Sensual experience could no longer be relied upon to confirm knowledge of truth since it varied from one person to another.

Unlike Plato, his senior by 44 years, Aristotle embraced the notion that reality did not preclude sensation. The distinction was a break in the tradition in believing existence and truth to include the property of changeless permanence, and consequently to the improper conclusion that knowledge *per se* required to be unchanging and permanent if it was to be acceptable as true.

Aristotle's theory of change held that three aspects are essential: a departing form, an arriving form and a continuing material presence. Change, Aristotle posited, is activated externally, requiring an agency or 'mover' other than that moved, on the premise that something cannot move itself. The 'mover' in turn may be moved by another mover, but ultimately the initiator of the first movement must be static.

Gross Change

The Greeks were apparently unwilling to entertain the idea that all could be in motion, and changeable. Their rationale led them to conclude that, at some point in the past, there must always have been a first movement. Aristotle ascribed such first movement to God, an authority that he had hitherto not acknowledged, presuming God not to be subject to change.

Net Change

Change may be said to be reticular. Its three-dimensional, net-like quality extends to envelop all in all directions, events being tied to all others by the fabric of space and time; and like space and time, nothing is free of its influence. Knots in the web are the simple means by which we articulate concepts, the strings their relations. If we have space and time, we also have the dynamic of change. Space was thought to be clearly differentiable until Hermann Minkowski tied it geometrically to time with a hyphen. The hyphen, seen or not, acknowledged or not, is the familial connection that confirms that any is a part of all. Space-time-change is a unity of interdependencies. That unity is all-inclusive. It is for the purpose of clarity of expression that we drop all but the most cogent hyphenated connections from our thoughts and communications, but they are all still there hanging-on-in-the-dark so to speak. The inventor-mathematician-designer-thinker-poet-author R. Buckminster Fuller had an infuriating writing style whereby he linked up to eight adjectives to a noun with hyphens, as if to emphasize that each was essential to the idea that he was trying to convey.

Unending Change

In our discussion of matter we postulated that the effect of heat was to facilitate change. If this holds true, then the corollary, that the effect of change is to facilitate the socialization of substance, follows. If one can accept this premise one can conjecture that in time all inclinations towards combination, including the most favourable conditions, will, on the basis of probability, occur. Nature can match all its parts to all its other parts in its unintended quest for affinities between all things and, in so doing, achieve equilibrium or equation between its dissimilar parts, if not simultaneously, then progressively in limited domains.

We now move a long way towards understanding and explaining the restless nature of nature, whereby its inventory of parts is 'searching' for its 'best' condition, that is to say, compatibility within its environment. This 'reality' can easily be recognized in the behaviour of people searching for fulfilment, which we call peace of mind, tranquility, knowledge of the rightness of their actions, compatibility within their environments, or unity in existence. Because we are burdened by will we are burdened by the need to establish our actions as purposeful, which purpose must be to seek unity in and of existence consistent with our natures.

All the positive and negative forces exhibited in human behaviour express the extent to which we have progressed in this direction, generally with a very poor understanding of our human nature. We find connections with our nature in our fragmental environments, in the effects upon us of music and other art forms, the wondrous beauty of natural landscapes, the excitement of learning of connections between hitherto unrelated subjects, the warmth that comes from witnessing human courage, compassion and tenderness. All these seemingly disconnected experiences have connections. The challenge is to search and discover how they fit together in the most plausible ways. The paths to harmony are tested by trial and error, a process that generates countless more mismatches than matches, mismatches that often carry a penalty of conflict.

Discordance is the prevailing characteristic of existence from which we draw our perspective of the paradox of duality. Whether reality is composed of opposites or our perception structures our experiences so, or both, our 'reality' is that of such a world. Why should opposites be so if not to serve as counterweights in an eternal self-correcting system wherein the predominant collective bias works towards the resolution of forces and unification, while the bias of each individual force is directed towards an extremity of experience and the promotion of imbalance?

The more that forces build in support of maintaining a state of imbalance, the greater the impetus builds behind the forces of change. It is as if, within any single system and set of circumstances, there is a point of optimal resolution. The pendulum is only momentarily at such a position, yet, on balance, it is always there.

Much as science has correctly assessed how food is converted to human energy, with great potential for good or evil, Einstein discovered in relativity the feeding processes of change. Given the seeming omnipotence of the forces for change, the ultimate test of durability or survivability is the extent to which an entity is responsive to change. Even the rich and powerful cannot insulate themselves against the vicissitudes of change. Our only recourse is to learn to adapt to its inexorable performance with open minds. This truism is applicable to the architecture of anything, and ultimately to the architecture of everything. In this way the universe of today must of necessity be other than what it was yesterday or will be tomorrow.

Life casts off that for which it has no further use and generates replacement tissues, resisting attrition of significant experience by engraving it in memory. Human society is endowed with a kind of corporate existence the effect of which is to desire to extend its best accomplishments into the future. The notion of irrevocable loss, as of a secret or virginity, leaves the individual with less integrity than it had, but generates a newfound integrity in that which it becomes. If there can be said to be a virtue in change, it is surely that it leaves open to possibility the prospects of change for the good, for integrity as wholes, and for integrity as probity.

Through this systemic processing of the equilibrium equation the two great universal tendencies, to continue doing and to modify doing, are maintained ad infinitum.

27   Exchange: Mixing and Matching

Nature, it is often assumed, is not concerned with efficiency. It generously generates great quantities of energy and matter of which minuscule quantities are applied to, for example, the direct support of life. But when this thesis is examined carefully it can be seen that each of Nature's elements recycles at different rates, and all its actors move on to new roles. It is the rate of moving on, that is change of doing, that can be found to reveal patterns of efficiency, averaged for each element and measurable by the number of times an element is utilized within a closed system before it drops out to perform another role within another system.

Let us take a quick look at the cyclical patterns of a few elements to show how they influence each other and propagate fluctuations in the world's atomic stock exchange. Later we will examine the composition of the atmosphere, ocean and soil to reveal distinctions in the flux of elements within these gaseous, liquid and solid environs.

Carbon

Free as air, carbon atoms, each united with two of oxygen, circumnavigate the earth as carbon dioxide until arrested by plants. Cuffed to hydrogen they are tried, stripped of energy and convicted to join the chain gang of countless biomass molecules sentenced to do forced labour as carbohydrates, proteins, nucleic acids, cellulose, chlorophyll, or lipids in the cells of roots, until discharged. There are other fates awaiting the carbon atom: being devoured by a microbe, intervention in the carbonic affairs of plants eaten by you, absorption in the oceans as bicarbonate or the slow decent through time as former sea shells and corals to become calcium carbonate rocks, each carbon atom with an expectation of reprieve and release somewhere, somehow, sometime, back into the air as free carbon dioxide until called upon once again to serve others. Carbon is therefore indigenous to each of the four primary pools: atmosphere, ocean, soil and the biota.

Hydrogen

Hydrogen, the lightest of the elements, rises into space and in consequence is present as a small proportion of the earth's atmosphere. Once released from captivity within the earth's crust in the form of gaseous volcanic emissions, through the degeneration of organic matter, or as a by-product of water vapour decomposed by ultraviolet sunlight, hydrogen can only retain its connection to the earth through alliance with other elements. Thus, hydrogen is dependent upon affinities that propagate changes to compound forms that render new properties and new relations. To this effect hydrogen is found naturally in alliance with carbon and oxygen in sugars, starches and cellulose, and in hydrocarbon forms that include fossil fuels of organic origin and in all acids, in which form it is present in almost all plant and animal life. It is through its facility to exchange or realign in partnership that hydrogen maintains its hold as the ninth most abundant element, comprising almost 1 per cent of the contents of the atmosphere, oceans and earth's crust combined.

Nitrogen

Nitrogen constitutes 78 per cent of air by volume. It keeps this steady state by reason of there being no other natural reservoirs in significant quantities, from which one might conclude that it is inert and useless. So it would be if it were not for the existence of certain bond-breaking bacteria with the faculty of converting nitrogen atoms into ammonium or nitrate. These maverick nitrifiers populate the surfaces of the oceans and attach to soil granules forming special relations with leguminous plants. Their unique contribution is to link nitrogen to oxygen in a form that can be accessed by other living organisms. Nitrogen has now passed from the atmosphere into the oceans and land, and as it penetrates deeper the concentrations of free oxygen get progressively lower. Life at these depths would suffocate for want of oxygen if it was not for the counter-forces of vast armies of subterranean anaerobic bacterial microbes, and navies of submarine microbial denitrifiers that treat nitrate as an oxidizer, utilizing its oxygen and releasing the nitrogen as waste. The free nitrogen rises to the surface and is released back into the atmosphere. Nitrogen cannot be said to be living by any criterion, but without its chemical aid much that is living would not.

Oxygen

Oxygen in all but trace amounts is the by-product of and life-giver to living systems. We shall address the emergence of oxygen and its relation to life in our discussion of life.

The accordance of significance to anything lies in the answer to the question: what does it do? Oxygen oxidizes. By this we mean that it facilitates chemical recomposition through combination with other elements. The term oxidation has been broadened in the light of findings that show that what is going on in such processes is the transference of electrons from one element to another resulting in the reversal of their polarities, as for example from sodium to chlorine to form sodium chloride where the sodium assumes a positive charge and chlorine a negative charge. Elements are said to be oxidized if they lose electrons, and reduced if they gain electrons.

The presence of oxygen is not a necessary precondition for oxidation but the involvement of oxygen in such processes is very extensive since oxygen can be made to combine directly with most elements, hence the term 'oxide' used to describe the products of such unions, which constitute the most common binary compounds in the earth's oceans and solid crust.

In this capacity oxygen functions as an accelerator of the global change mechanism. The presence of oxygen is the reason why the earth exhibits the rates of change that it does, while the moon relies almost entirely upon external forces to bring about change. Impressions of the impact of meteors and astronauts' boots and scoops are all the changes that the moon has to show for millions of years of presence.

Elements

As we have seen, elements cycle separately, but in their respective processes some find it advantageous to work in unison with others. Thus, for example, nitrogen and phosphorus, the bulk of which predominate in the realms of atmosphere and soil respectively, hinge inextricably together in the oceans to work in the ratio of 7:1. Deviations from the set ratio trigger a switch from the function of linking to that of disengaging nitrogen and phosphorus, as the conditions dictate. The connection between this rational phenomenon and life is a subliminal sub-living relation by which the same ratio may be found in the organic oceanic floaters as in human life, a symbiotic relation where the organic and inorganic work together for the benefit of each and of the whole of which they are parts.

Compounds

A multitude of compounds comprises only hydrogen and carbon; called hydrocarbons, they differ only in their constituent proportions. Methane and butane are but two examples.

The duo of calcium and phosphorus are worthy of comparison. Calcium passes through the plant/soil life cycle on a one-time use basis, draining off in soluble forms to rivers. The supply of calcium for replenishment is bountiful. Phosphorus, being about one-40th as plentiful as calcium, has a proportionately greater economic value to plant life. By coincidence, divine providence or mathematical probability, a single atom of phosphorus, being less soluble than calcium, will hang around and recycle on average forty-plus times through the symbiotic plant/soil menu before passing on to less green pastures. Phosphorus happens, nevertheless, to place critical nutrient limitations on the abundance of plant life. By such comparisons we can observe how nature reconciles her needs to her supplies, salvaging and recycling the less abundant nutrients to the effect of enhancing the total generation of life forms towards its maximum potential, given the finite supplies available. Thus, the eddy counts of individual constituents contribute to the whirlpools that comprise the global exchange mechanism.

We will examine the teamwork of carbon and oxygen coupled as carbon dioxide when we address life. Carbon dioxide has the distinctive properties of permitting the passage of short-wave energy represented by sunlight while blocking the path of long-wave infrared energy characteristic of heat emissions from earth. As such, it played a vital role in ameliorating conditions favourable to the establishment of life. In the tide of evolution elements have bonded to become compounds, compounds compounded to become organisms and organisms, in ever-increasing numbers and variety, have become more specialized and sophisticated. Variations, subscribing to evolution, are identifiable backward in time and towards a time when there were many fewer varieties, and ultimately in logic to a distant time when there was but a single species.

The Stock Exchange

The effect of the mill of exchange is therefore to enhance probabilities that ambient conditions, critical masses and mixes will, some time, somewhere, likely occur that prompt the formation of life and offer it prospects for continuance and development. The big league players are not so by virtue of their preponderance in numbers but for the critical roles that they play. Many of the elements occur in approximately the same proportions in all live things, a fact that efficiently rewards eating.

The effect of exchange in nature, and the purpose of exchange in human consciousness, is to facilitate matching. Only through circulation can probability of acquaintance be increased. Only through acquaintance can the test of compatibility be made. Only through compatibility can suitability or fittingness be confirmed. Only through fittingness can unity be achieved. Such is the mix and match prescription. In failure to achieve compatibility the maintenance of separation serves to stir the atomic caldron further and in so doing enhances the probability of a subsequent match. All this sounds implausibly close to the exercise of will. Perhaps we can rest more comfortably entertaining the notion of fate. What else can we understand is going on that is exhibited through the auspices of exchange?

Self-Regulated Change

Exchange places limits on continuity. Exchange ensures that nothing and no system can over-extend its allotted time, yet all appears to be not merely possible but probable! Metaphorically, all combinations of matter appear to be destined to be, and not to be, 'mortal', having time-stamped opportunities to come into being, exist, corrupt and die.

So, sooner or later, why should mortals not walk and talk?

28   Event: The Happening of a Change Phenomenon

An event is the occurrence of a phenomenon or complexity of process that is confined in time between a starting point and an ending, and limited in spatial extent.

Ambivalence

If we refer to a dictionary to establish meaning for the word 'event' we find that it has more than one, which is a common enough experience, but that such meanings are contradictory is not. An event defined as a change or succession of changes differs from an event defined as the product or consequence of change. It is easy to infer that each is referring to a different state of change but confusion of the two is understandable. To which each 'event' refers is not always explicitly stated, from which uncertainty we have difficulty in establishing which definition is intended to apply. We can now see how confusion may arise with respect to how sequence as precedence, concurrence and consequence determine the nature of subsequent definitions. The contrary nature arises out of a confusion of time.

Underlying any such confusion is the fact that events compound upon each other, having occurred in the past, occurring in the present and awaiting occurrence in the future. While all this seems to correspond to our understanding of sequence in time, we also understand that all events occur in the present and therefore cannot occur in the future.

From our sloppy usage of the term 'event' in this way we confuse the idea 'event' with the idea 'advent', being any coming or arrival, distinguished from a happening that may extend beyond its initiation. Events are being confused with their relation to other events. It is the relation of cause to effect that we must look at to better understand the nature of specific events.

Clarity

The nub of an event is neither the cause nor the effect of that event, but the incidence of change. Thus, the significance of lightning striking a tree lies not in the passage of energy from the atmosphere to the earth, nor in the fact that a forest may burn up or down, but that those two events share a momentary connection in time and place when the lightning has left the sky but before the fire has started, which unique event would be different had it occurred in another place at another time. An event is singular.

Remembering, as we addressed the subject of time, that the past and future were connected at a point, 'now', and that 'now' appears to move away from the past and towards the future, and having since discussed change as being a break in continuity, there are within these notions segments of time, evident at particular locations, where a change is followed by continuity until interrupted by another change. Such a span of continuity, however long or brief, can, by some accounts, also be characterized as an event, even if no change occurs within a set time frame. Thus, an event must have a beginning, duration and an end. Within an event any number of shorter events may occur, concurrently or in sequence. An event must therefore be qualified by language to distinguish it from any others.

Insofar as an event must occur in time and space, its correct description should tie it to each as a space-time-event, by which means we connect what is occurring to the where and when of its occurrence. For an answer to the question 'why' we look for a preceding cause, and thence for an answer to the question 'how'. We understand that when the 'what', 'where', 'why' and 'when' questions are composed, and answers disclosed, we are led through the effect to an understanding of 'how' that is manifest in the event itself, the happening of the change phenomenon, and of the form that the effect assumes.

29 Cause and Effect: The Pecking Order of Influence

Events, as we have shown, tend to fit into one of three time frames: the past, the present or the future. If we look at a series of three events relating to each other in sequence, we tend to focus upon that event that is occurring in the present, disregarding that which preceded it, and giving scant consideration to the one that will immediately follow. What is common to each of the classes of events is that each preconditions the one that follows, enabling, limiting or denying later possibilities by their nature. 'Ifs' cause 'thens'!

An event that contributes to or dictates events that follow we call a cause, while an event that is the product of such prior facilitation we call an effect. A cause is an event in precedence, an effect an event in consequence. An event is thus 'born' as an effect of earlier events and 'dies' as a cause of later events.

First Cause

The first problem that we have with this rationale is that we are not willing to accept that the chain of events could possibly have proceeded from an infinite past, or to accept that it will continue into an infinite future. We have thus set ourselves an unnecessary problem that we are unable to solve, namely the problem of accounting for the first cause. While divine providence may offer a satisfying alternative to an infinite past, this idea denies access to any accounting for first cause of the divinity itself, thereby failing to offer a fully satisfactory explanation of origination.

How, we may ask, do we get into such intractable quandaries? To such questions we invariably prefix our answers with 'because', meaning 'by cause' or due to cause.

The chicken and the egg dichotomy, i.e. which came first, is not exactly as it may appear. This dilemma arises out of our inability to distinguish any essential difference between one chicken and another, or one egg and another through the course of our life spans. What we tend to disregard is that one human life span is but a flicker in the span of time that life has taken to establish itself on earth, and that, through the aegis of evolution, or any other accounting system one may choose to substitute, chickens and chicken eggs have developed from simpler life and life-bearing forms, all changes to which are attributable to causes and effects. A chicken is an evolved egg, an egg an extension of chicken.

Continuity

It is impossible to dissociate an effect from being a cause except as it may relate to a specific event. By analogy a cause may be likened to energy that affects matter, while fully understanding that both are both. The noun 'cause' is descriptive of the actor, the verb 'cause' of the action.

We should acknowledge that, for practical purposes, a cause is only so called if it is sufficiently influential to be credited as contributing to an experienced effect. As to whether the aforementioned Peking butterfly is sufficiently influential to affect the weather in New York depends upon the definition of sufficiency. Unless the debate is conclusively proven one way or another, nobody can afford to claim certainty in matters of degree of any contribution to any cause. Thus, we may choose to ignore the butterfly, or give it the benefit of the doubt. In so doing and for our current purpose we define arbitrary thresholds of sufficiency, influence, credibility and contribution.

Wilful Causes

A seemingly necessary condition that relates a cause to an effect is connection. There must be some connection that transforms the radiating influence of a cause to become the illuminating influence upon an effect. Such connections are called events.

Human initiative may be seen as a cause of causes, as wilful induction, where action is attributable to human judgement and decision. It has a special place in the relation of events, as of privilege, weighting in favour of or against actions that would otherwise take place in nature without our intervention.

This idea of causes of causes may be compounded by human design to the degree that man can step aside from a sequence of events in the expectation that his will and intervention may be fulfilled automatically once he has set his agents in place and initiated their performance. Mechanical chain reactions and the so-called domino effect come to mind. The converse also applies where failure to intervene wreaks havoc upon otherwise well-laid plans. For want of a nail a shoe is lost, for want of a shoe a horse is lost, for want of a horse a rider is lost, for want of a rider a battle is lost, and for want of a battle a war is lost! Butterflies and nails share common ground 'because they be causes'.

It is generally assumed that a 'causer' acts with a greater degree of free will than that which is acted upon; however, that conclusion is too discrete to fully account for the action itself. The 'causer' is under similar constraints to those conditioning the 'causee', being beset by prior causes, and as such cannot be said to be any more fully in control of its destiny, except as it relates to events immediately following. As to whether such prior constraints are sufficient grounds to excuse the actions of the 'causer' is the subject of more than 2,000 years of legal debate. As wilful animals we have a greater measure of control over the effects of our actions than we have over the causes. As a result we bear a greater burden of responsibility than animals and are rewarded or punished proportionately for our good and bad behaviour.

Confusion

Cause appears to derive from cause as a necessity, but is this necessarily so? Could not effect be as equally attributed to a 'pull' as to a 'push'? Because we associate two bodies with giving appearances of gravitational attraction towards each other, do we not assume that, since all similar pairs of bodies appear to behave in similar ways, that they are responding to similar forces of attraction? Have we not drawn our conclusion of today's event on the basis of yesterday's and in so doing established attraction as being factual? How can we be sure that the opposing condition is not governing whereby the two are behaving according to the principle of repulsion whereby they abhor the antithesis of matter: vacuum? Perhaps some are motivated by a 'push' while others are motivated by a 'pull', as by an alternating current of forces! Doesn't the whole idea of an alternating current obfuscate the question as to what is pushing or pulling what, if they both do both?

Clarification

If we look to the immediate antecedent event as the probable cause we are greatly simplifying the complexity of events that have led up to and contributed to a succeeding event. What we are doing, often without acknowledging so, is establishing in our minds a 'primary cause', which, because of its immediacy to the event under consideration, often appears to be that cause which contributes the most forceful influence upon a subsequent event. In the case where hereditary traits are passed from parent to child, the primary cause is the parent, but as to whether the burdens of responsibility and accountability should be placed on the parent is open to question since the parent is simply a link in a chain of succession that has until recently, and now only in a few cases, lain beyond all prospects of personal knowledge in the clinical sense. So a cause that we identify to be such is simply one particular one of many, identified in our minds as worthy of our attention in exercising free will.

In everyday language the term 'cause' continues to carry the connotation of principle contribution to an event, there being no need to analyse and account for all contributions, or to make distinctions between induction and explanation to which we have referred. We can now recognize that cause is not, except in the case of the human, something that is consciously intended to be and do, but rather something that is responding to a law or laws of nature. In this interpretation responsibility for action is removed from the object of doing, as a falling stone, and of necessity transferred to the agency of enforcement, as the force of gravity. Now science can endorse such knowledge in the form of general principles without being concerned with every specific and immediate instrument and instance of cause.

The only exception to the unconscious rule of law is human. The human, through utilizing perception and exercising judgement, has expanded the range of possible effects that can result from a particular set of causes through his capacity to make decisions, even radical decisions, and act upon them. Why would nature allow such a potentially disastrous course of development to occur? The short answer is that nature has made no exception for human behaviour. Nature's legitimate interests are served by man in the same way as they are served by other animals, by allowing them to make decisions, conscious or otherwise, in their own perceived best interests, while maintaining them subject to her cardinal rules of the universe. In the case of emerging intelligence in man, while individual men may attempt to act in ways contrary to the laws of nature, by the laws of probability most will not and the survival prospects of the many will far outnumber those of the few, thus maintaining a state of general compliance with the laws.

We should not leave this subject without noting a mistake that we are inclined to make; the substitution of the idea of cause as 'explanation' for the idea of cause as 'motivation'. A description is merely a verbal painting of an impression that we have of a state of relations at a particular time and place, whereas motivation is a compulsive force that must be recognized as belonging to the body of law we call physics. The distinction is that between quality and quantity.

We should recognize that in order to induce a specific effect it is not necessary to precede that event with a singular and specific cause. $4 + 0$ will produce 4 (the inertial model), but so will $3 + 1$ or $2 + 2$. There are, in fact, an infinite number of paths that a comedian may take to induce a single laugh, which realization helps to explain why we can often agree upon ends while differing and dickering over the most effective means.

The point at which a cause becomes an effect is relative to an observer or other point in space and time, as may be illustrated in comparing the times that a flash of lightning and a clap of thunder take to reach an observer one mile from an event with the times they take to reach a second observer ten miles from the same event. The speeds of light and sound transmitted over different distances determine the earliest moments that the causes appear to take effect from different viewpoints. Thus, causes and effects become personalized much as we have already discovered do time and space, perception becoming indistinguishable from imperception, and subject from object!

We shall leave this subject with the parting suggestion that in looking for the line of succession of causes-of-causes as far back as can reasonably be traced (rather than just looking for and accepting immediate causes) is fundamental to overcoming many of mankind's most obdurate problems. In doing so we are transported back into a childlike innocence whereby every answer that we receive stimulates a further question 'why?': Such a process should prompt us to wonder what we are losing from childhood as we grow older but not always wiser.

30   Productivity: Intensity of Doing

Performance as the act of doing, or in hindsight that which has been done, is a process or product of the expression of energy through the direction of force. In using the word 'productivity' we generally have in mind a sense of achievement which, consciously or not, we compare with similar events in our experience. Such comparison often lends itself to formal scales of measurement, as when we measure the efficiency of something as a ratio of the input of energy to the output of work. In this way and over time we can represent similar experiences on a scale ranging from the lowest to the highest degree of effectiveness, which scale we can then use as a basis for assessing the performance of later experiences.

Individual vs Collective Performance

Two distinctly differing concepts of performance arise: that which is the capacity for doing of a single entity, as a lever purchasing over a fulcrum, a single-issue event; and that which, due to the interaction of many parts, achieves an effect that a lesser number of parts cannot achieve.

In the former concept we assess performance in terms of the degree to which simple efficiency is achieved, the lever arms being designed to take into account and reconcile the force required to lift a load with that available to lift it. We measure the efficiency of such simple mechanisms as the ratio of the two forces and the lengths of the two lever arms recompense those proportions.

In the second example, most of the parts will likely not be performing at their optimal, most efficient or productive rates but will contribute what they do in proportion to what the total entity requires of them for it to function as a whole. The optimal performance of a part may be compromised for the optimal performance of the whole.

Ergonomics and Satisfaction

When man engages in the design of complex organisms he does so in the context of his own efficiency, i.e. as a ratio of cost to benefit, the efficiency of the working parts and the efficiency of the total organism. Under these governances he will opt for the use of parts that, when working together, approach their optimal potential simultaneously. That is the most efficient model for design and for the production of benefit. In a machine such efficiencies would be reached when all parts approach their yield points simultaneously and, if further stressed, would all fail at the same time, exceeding the limitations of their designs and physical strength.

However, when we gauge performance as a measure of satisfaction, we should be aware that we are no longer gauging the performance of something or some compound thing by its capacity to do, but rather by its contribution to the generation of pleasure on a comparative but non-scientific level.

Peak Performance vs. Maintenance

Another criterion by which we assess performance is through measuring the ability of something to peak at its highest level of performance compared with an ability to sustain a high level of performance. We recognize this distinction when we compare the relative performances of two athletes, say the high jumper with the long distance runner, or two mechanisms, the hammer with the wheel, the performance of each of which can be further enhanced with practice until physical limits are reached, complacency or habit take over and performance is repeated by rote.

It can readily be recognized that the scales appropriate for measuring performance will differ as suggested by each enterprise. Thus, enhanced performance of the high jumper is assessed on an ascending scale while that of the long distance runner is assessed on a descending scale.

31   Attribution: Intrinsic or Ascriptive Assignment

When we examined those qualities or quantities that were essential to existence we accorded them a state of being intrinsic, by which we meant that something could not be what it is or was in their absence. These qualities or quantities we attributed to things as essences, the presence of which confirmed that things were what we believed them to be rather than something else. By such determinations we secure foundations to our definitions. By our own definitions things are what we say they are, even if we claim such fundaments to be intrinsic to the things themselves.

We are the instruments of association of a thing with its essence. We attribute intrinsicality to a thing as a matter of fact because we accept it as such as a matter of definition. About this we can be certain.

Where we are less than certain that a connection exists between two entities but suspect that one depends upon the presence of the other, and in the absence of better information, we ascribe the qualities or quantities of the one to the presence of the other, until better information gives us reason to reconsider their relations. Knowledge is in this regard more or less secure.

What then, we may reasonably ask, is the difference between according and believing? The act of according formally sanctions our understanding of relations based upon our judgements, for our convenience. Both are the same insofar as they rely for their existence and distinctions upon our acts of intrinsic or ascriptive assignment. Attribution is the act of according the property or properties of one thing the status of being of necessity dependent upon another, as effect depends upon cause.

32 Chance: The Absence of Assignable Cause

Having carefully determined that all is attributable to cause, what accounts for the maverick experience whereby an event presents itself as a product of indetermination? The quick answer is 'appearances'. What is lacking is evidence as to what the determinants are. In the absence of knowledge enabling attribution we either presume, or we elect not to presume. When we presume, we gamble on a hunch that events will bear out that our intuition is correct. When we decline to presume, we accept that events will transpire regardless of our understanding. Gambling is a game of chance. Casino operators don't gamble, they have knowledge of probabilities. Probabilities are not so by chance. It is the clients of casinos that gamble for reasons that have little to do with probabilities of winning.

To admit chance, we might say, is entertaining a mix of probabilities and improbabilities. We are all only too aware that events occur all the time that are beyond our powers of anticipation and control, but the idea that we might allow events to make decisions on our behalf when we have alternative and more secure options, many would say is the exercise of poor judgement at best and the abrogation of judgement at worst. Such casual concern for the quality of decision-making is the arbitrary negation of discretion in favour of unrestrained random selection. In wilfully admitting random selection when more promising alternatives are at hand, we become naked thrill-seekers enhancing the probability and inviting the thrill of a spill.

Cause is the strongest bearing force. Insofar as all effects are the products of cause, the certainty of cause, not to be confused with the absence of choice, translates to the absence of chance. If assignable cause is cause for consideration and the product of decision, nothing is left to chance except that which is a negation of sound judgement, because that is how we define chance.

Distinctions between chance as randomness, necessity as a normal manifestation of nature, and purposeful design are simply expressions of probability predicated upon differing degrees of possibility.

33 Power: Relative Influence

The common thread that extends through the passage of thought from cause to effect to performance to attribute is the inherent property of power. By power we mean potential or ability to exert influence; to do, act or to perform work.

Our general understanding of work carries a connotation of affecting change. How much work can be performed, or the extent that change may be effected in a measure of time without consideration for the factor of speed, is the measure of power. When we are exposed to the idea of power two forms present themselves: physical power, in which category the potential of a dammed reservoir belongs, and mental power, which embraces the potentials of the intellect, volition and the affections.

Since we have broached the former subject in addressing energy, matter and their relations, and we have yet to address the latter subjects, we shall limit our attention here to a few general principles. Simply stated and as a prelude to examining the operation of power, a few truisms attend the fact of power.

In general, power governs the forces that direct events that determine futures. As such power is an exponential expression of energy. Such power is value-neutral, being quantitative rather than qualitative. Couched in other concepts, as might, strength, capacity, authority or dominion, power 'lies, taut yet living, coiled, the spring', to quote Jacob Bronowski from *The Abacus and the Rose.*

In particular, personal power is perceived as security against adversities. It is a psychological path to and from authority through which a meaning to life can be educed.

34 Force: Energy Acting in a Cause

As we have stated, force is the release and exercise of the potential of energy in a direction. First, we approach this subject from a position of ignorance and therefore innocence. Through accrual of experience we become familiar with force, forms by which it is delivered, and with the effects of its impact. To segregate and articulate the essence of force has taken centuries. Concepts have been tested and found wanting, and new theories have supplanted them. It has been, and will continue to be, a process of building a body of knowledge subject to greater enlightenment.

Confining ourselves to the discussion of types of force-bearing particles found in nature, distinguished from types as they may be classified by agency or mode that are discussed as examples elsewhere, there are but four. They are, listed in their ascending order of intensity: gravity, the weak nuclear force, electromagnetism and the strong nuclear force.

Gravity

The first and most familiar of these forces is gravity. Gravity is the weakest of the four forces in that it is the easiest to resist. Even a fly can resist the forces of gravity for a short period. Gravity affects all matter that possesses mass, from the particle to the star, such that each appears to attract all others in proportion to its mass, conditioned, as Newton discovered, by the square of their distance apart.

The great surprise that attends gravity, of which Galileo was aware, is that it was found to act on all bodies in the same way regardless of their masses. Specifically, this translates to reveal that, if the distance between two bodies increases by a factor of two, the strength of the gravitational forces between them decreases by a factor of four. If this were to happen in a limited field, they would all pull into the centre and consolidate. For this not to occur suggests either a compensating outward movement of all matter away from a central point as would be occasioned by an explosion, or a distribution of matter approximately equal in density and unlimited in extent such that each body would have equal forces acting upon it from all directions.

Alternatively, the alternation of explosions and implosions in such a way that the total volume of infinity would not change but rather comprise a dynamic suggestive of a community of breathers, all respiring out of synchrony but on average maintaining a uniform density. We may get into trouble trying to assign polarities to each body of matter in pursuing this idea, but the problem of polarity would be overcome if the current and therefore the polarity were alternating. In our experience all phenomena have their polar opposites and seem to exert influence upon each other in some reciprocal pattern. So it would be rather surprising if we were to discover a phenomenon (say the universe) that did not behave in such an alternating manner.

The Big Bang Theory to which we have earlier referred illustrates the notion of an outwardly expanding universe that not only offsets gravitational forces that otherwise contribute to an implosion, but over-acts to further separate bodies by applying excessive explosive force. If we maintain the position that gravity continues to draw all bodies together, we have necessarily to raise the question as to whether, in time, the effect will be to exhaust the forces of universal expansion. By implication this model signals finite limits for both expanding and contracting universes as inevitable consequences.

The counter-argument would suggest that, while gravitational forces acting between two bodies diminishes in an expanding universe, the amount of force required to continue the outward journey is diminished in exactly the same proportion by the same rule, thus favouring infinite expansion! If the force of arguments in support of a limited, as opposed to a limitless universe are equally compelling, then we have in effect a steady state.

While a steady state could theoretically be maintained by approximately equal inward and outward forces acting to compensate each other, such a system would depend upon a delicate balance between the two. The maintenance of such a balance would be problematic, given the degree of movement observable in the universe, in the absence of some overall system of control. Any control system would need to accommodate changing relations between bodies. Such linkages would need to resist compressive or tensile forces to maintain approximate equilibrium.

In principle such universal connections are analogous to a system of invisible rubber bands whereby all bodies are connected to all other bodies. As one body is drawn towards another, those bands on its opposite side come under greater tension, while the bands on the side facing the second body slacken. The net effect would be for the first body to draw back to a neutral position where all forces bearing upon it would be approximately equal. If one body was moving relative to any other, the network of rubber bands would tend to draw all others into motion, as if in a universal quiver, self-correcting any tendencies in its members to stray. If this is indeed so, then why, we may ask, do meteors crash on to the surfaces of planets? To which a plausible answer would be for the same reason that delinquents get into trouble in our own experience, because they unduly stretch the rules, deviate too far from established norms and come under the dominant influence of a force more powerful than the sum total of all forces acting to keep them from doing so. Thus, the larger bodies serve as celestial vacuum cleaners consolidating misadventurous cosmic jetsam. The question then arises as to the nature of the critical threshold that determines whether a small body is captured by a larger body or not. The answer would appear to be a function of their relative masses, velocities and directions combined with the dynamic influence of all other bodies bearing upon them.

An attractive aspect of this rubber-band theory is that the total system of relations is not so delicately contingent upon balance as may otherwise be inferred. As to whether the rubber-band theory is not simply a representation of the steady-state theory, of the concept 'aether' used to account for action-at-a-distance (discussed later), or even of relativity, we shall leave to those with professional advantage, and to later sections on belief, knowledge, theory and symbolism.

If the primary fact of nature is rubber-like, we do not have to worry about elections between the left and right extremes of politics stretching to exaggerate their positions, for there will always be a rebound into the opposite court. It is only limitations of personal vision and self-interest that cause us to worry about short-term effects like having more or less of anything, including life. Rest in peace, the game will continue.

The force of gravity is paradoxical in that, while the mutual attraction between bodies may be confirmed by observation and measurement, no direct connection between such bodies is detectable. All that is needed is a linear relation. Another analogy of this relationship is suggestive of an electromagnet which, if applied to this situation, would treat the two bodies as being on a common axis and functioning as extremities of a central core. The gravitational pull of all other bodies off this alignment would form an electromagnetic field, which would induce magnetism between the two bodies in question, polarizing one as positive and the other as negative, and thereby account for the attraction between the two bodies in question. The respective masses of the bodies and their distance apart would, according to Newton's second law of motion, determine the velocity and acceleration by which each is drawn towards the other. When the velocity of their convergence is measured it reads as a single number and masks what is believed to be a composite of two separate forces. That belief is founded, perhaps erroneously, on the observed condition of bodies appearing to attract, as apples falling to earth, whereas no such attraction would exist if the apple did not exist, from which one might conclude that gravity *per se* does not exist as a property of matter, but rather of relations, or more succinctly, of both.

Returning to the strange subject of spin for a moment, as we have noted, all matter embodies spin 1/2. The relations between distanced matter governed by gravity are attributed to a quality, spin 2, that is resident in a particle and called a graviton. If gravitons may be likened to a stream of energy much as light links the sun to the earth, then we may reconsider the idea that there may be direct connections between such bodies.

The Weak Nuclear Force

The weak nuclear force is that which interrelates hadrons with leptons. Hadrons and leptons are not particles but rather families to which certain types of particles belong. For example, the protons and neutrons forming the nucleus of the atom belong to the baryon subfamily within the hadron family, while the electron of the atom belongs to the lepton family. So the weak force describes or rather defines the relations between the two.

The weak nuclear force operates at frequencies of one metre and more within the band range of radio waves and acts upon all matter of spin 1/2 (but not of spin 0, 1 or 2) to cohere some classes of matter and not others, as the positively charged nucleus of one atom may be attracted to the negatively charged electron of another. At the molecular level of compounds, similarly, the weak force will draw the positively charged electron of one atom of one molecule, towards the negatively charged nucleus of an atom of another molecule, as when two molecules of water bond negative to positive ends in an arrangement illustrated by the numeral 69. The principle function of the weak nuclear force is to bind atoms to atoms, atoms to molecules and, therefore, molecules to molecules.

Electromagnetism

The third force we shall consider is electromagnetism. Electromagnetism is the broad spectrum of magnetic conditions that result from currents of electricity interacting between electrically charged particles, as between positively charged protons and neutrons, and negatively charged electrons. This attraction is attributed to an exchange of massless particles called photons, with the attribute spin 1.

The degree of excitation of atoms by energy is inversely proportional to the length of the wave; thus, radio waves in the low frequency bands do not readily induce heat, while microwaves are more effective, and visible light, X-ray and gamma rays at the high-frequency end of the electromagnetic spectrum, which contain greater amounts of energy, have a higher potential for conversion to heat.

Electromagnetism binds electrons to the nucleus within the atom.

The Strong Nuclear Force

The strong nuclear force is that which binds the parts of parts of the atom's nucleus, known as quarks, to each other. Since quarks constitute both the hadron components, the positive protons and the uncharged neutrons, quarks serve to bind the whole nucleus into a unit. The relations of such entities within the atom that are governed by the strong nuclear force are believed to be controlled by gluons, particles with the quality spin 1, that only attract their own kind, gluons. We addressed sub-atomic particles in our discussions of matter.

A particular quality of the quark found in the 'strong nuclear force' is that the gluons that bind the quarks together bear a peculiar quality of their own known as 'color'. This quality, 'color', bearing no resemblance to colour as we customarily understand it, has the distinctive property of not diminishing with distance as Newton would suggest and is the case with electromagnetism, but increases with distance, thus supporting the aforementioned rubber-band theory wherein the tension increases with elongation. The effect is to reinforce the attraction of quark to quark within a hadron.

In General

The delicacy of the relation of the four forces lies in their hierarchy. Acting within narrow ranges of variation, the effects of one do not unduly disturb another. Were any one to become significantly stronger or weaker, the whole structure of the hierarchy would falter. If gluons became unglued, for example, that is, if the strong nuclear force became less strong, the next force in strength, electromagnetism, would become relatively more influential and deny the unification now achieved by the strong nuclear force, in turn denying the structuring of atoms and the existence of elements. In a nutshell, the hierarchy plays out on the basis of the rule 'the closer the proximity the tighter the bond'. Such is family life!

Force is the active agent of cause that produces an effect, the degree of which is determined as a function of mass multiplied by acceleration measurable in newtons. The newton is simply a quantifiable amount of mass subjected to a known acceleration, one newton being equal to one kilogram of mass multiplied by an acceleration of one metre per second per second.

Weight (measurable effect of gravity) can be argued to be an inconstant by virtue of the counter-gravitational forces of the moon, planets, sun and stars changing their directions and distances relative to and from any object under consideration. What we value in according weight to an entity therefore is its rank in the hierarchy of other weights so that we can make useful predictions about their respective behaviours. This 'rank' does not vary at any given place and time.

35   Development: Power in Operation

For the purposes of this discussion the term development is used to embrace processes of change in the arrangement and attributes of entities, and particularly the manner by which they form larger entities.

We distinguish between arrangements resulting from formation, deposition or growth attributable to causes other than conscious acts, and those that are the products of deliberation and decision. The reader will be quick to recognize that the boundary between the two is a little fuzzy, within which fuzz the animal kingdom grows and constructs its shelters. For this reason we attempt to offer a concise definition of consciousness in a later section.

We have discussed the bonding of sub-atomic particles to form atoms, the coherence of atoms to form molecules, the aggregation or congregation of like kinds to form enlarged bodies of matter, and have alluded to and will subsequently further develop the processes of change and exchange that result in hybrid mixes and new associations. This we call by another name, architecture, spelt with a small 'a' but accorded a large meaning. By this name we signify formation, a word like 'development' that is used as a verb to signify process, and as a noun to denote the product of such a process, syntax determining which meaning is intended.

Generally, we apply the term development to the organization of inorganic, elementary matter. The narrower context of development as it applies to the growth of living things is treated after our introduction of 'life'. The more limited meaning of development, that which is the result of conscious human intervention in the otherwise impersonal world, the application of creative work and its productive results by intention, we shall postpone awhile and address as aspects of art, science, industry and design, where the subject of the architecture of man will arise again.

Development in its broadest meaning embraces the processes and products of change that encompass the assemblage of elements and the disassembly, breakdown or simplification of relations between parts. In this capacity the term 'development' is synonymous with change.

36   Creativity: Expressions of the Maker

During the 'seven days' of creation, unconstrained by time, space, matter and energy, God created the universe and mankind, of whom, today, one-seventh are born on the Sabbath, one-sixth are Muslims, one-fifth are Chinese, one-quarter are children, one third are under-nourished, half are women and all are mortal. What more can be said about the agony of creativity? We shall suspend further discussion of creativity that arises due to alleged wilful acts of God (about which nobody can speak with authority) for inclusion in Section 205 on Theism.

We will also limit discussion of creativity to the production of novelty excluding the processes of human intellect, volition and the affections, under which headings considerations regarding human creativity arise in the abstract in imagination and are fulfilled through decisions, work and devotion. Creativity as activity resulting in newness is as commonplace as action itself in that it results in the creation of new relations.

There is in this broad definition no necessity to invoke purpose, will or intent. Creation happens as the direct consequence of unequal and conflicting forces. Force and direction as makers create novel expressions of force and direction. Similarly, no assessment can be placed upon that so created that establishes it as having any intrinsic value. A creation is in effect an effect.

Creation does not need to be tangible or sensible or even capable of discovery; indeed, the less accessible it is, the more we tend to revere its presumed source and cause. Magicians and ministers of all persuasions depend upon professional mystique and invisible sources for their authority and livelihoods. In this sense both creatively exploit ignorance and innocence.

How can we describe the creation of a novelty without referring to our inventories of near-like experience? Thesauri with hundreds of thousands of synonyms and antonyms have been compiled for the express purpose of offering a selection of words from which to draw precedents that most closely match or mirror the novelties that we are trying to describe. Beyond the harnessing of like words to describe our thoughts, we resort to the composition of metaphors and similes the number of which is essentially unlimited. What we understand from this is that the fact of creation, much like metaphors and similes themselves, must of necessity involve the arrangement of parts that are already created.

Because we are human we tend to attribute creation to human faculties with which we are familiar, even when we know that humanity has made no contribution to such creations. We can respond well to, identify with and therefore accept creativity more easily when we ascribe doing to a doer. Connotations of will, purpose, decision, work and the commitments that they imply underscore the work of doers. In their absence, acts of creation are more difficult to comprehend and accept. From such premises it is a lesser step to accord tribute to the maker of works and reverence for the genius that is presumed to lie behind works of extraordinary ingenuity and beauty. Both the attribution and the accordance advance the cause of authenticity of our suspicions that somehow, somewhere, something or someone is responsible.

37   Inertia: Predisposition Towards Continuance

We observe the tendency of static bodies to remain static and moving bodies to continue moving in the same direction at a constant velocity until acted upon by external forces. This tendency is called inertia, the principles of which were first established intuitively and later confirmed by experiment by Galileo Galilei. One might almost suppose that such bodies 'want' to continue what they are doing peacefully, being themselves and minding their own business.

Universal Verities

The most controlling criterion that governs our access to knowledge of the universe is the speed of light. As a body accelerates it acquires greater mass. At the speed of light it acquires infinite mass or resistance to further acceleration, thus establishing an absolute maximum velocity. The only reason that light can travel at the speed that it does is because its particles, photons, are massless and, as a result, will not increase in mass exponentially with an increase in velocity.

If our universe is expanding it is doing so at a fractional rate of the speed of light, a rate sufficient to overcome all gravitational forces that act to draw it together, yet not so great as to attain the speed of light, at which point the mass of matter in the universe would increase infinitely, thus requiring an infinite amount of energy to further accelerate. We do not know the rate of expansion of the universe for reasons that we shall discover when we examine the concept of uncertainty, but in essence this is because we cannot simultaneously focus upon place and time with a high degree of accuracy.

Why then does sight, receiving signals at the speed of light, not extend beyond the boundaries of the universe that are receding at a slower rate? One explanation suggests that if there is no source of light outside our universe, then we cannot sense any such presence. A second consideration might be that light passing from the boundary of the universe to our eyes would require to offset the effects of increases in distance due to expansion, increases that double when we attempt to view parts of the universe beyond any theoretical centre point. The passage of light from the boundaries of an expanding universe to our eyes takes light-years, so, even if we were to detect light emanating from such boundaries, it would only tell us about conditions that existed light-years ago. The boundaries will have extended much further from us by the date of our observation. Even light travelling at the maximum speed possible takes time. It is this passage of time that ensures that we see everything in the past and nothing truly in present time.

The Immutable Push-Me-Pull-You

The self-generated addition of mass due to acceleration is the governor that puts the brakes on further acceleration. Diminishing mass proportionate to deceleration is the control mechanism that ensures that velocity is maintained by exactly equating force with resistance. Equilibrium results from forces being offset by equal counter-forces.

Inertia is therefore a self-fulfilling phenomenon. It is the product of not being able to accelerate within a constant friction-free medium without the aid of additional force, and not being able to slow down in the same medium without acquiring additional resistance, because doing so will reduce its mass and therefore enable it to accelerate. Remember that mass is the quantity of matter in a body, inertia its resistance to acceleration.

Being Sensible

Stephen Hawking tells us that if light cannot pass from one region to another, no other information can. If so, we can only speculate and imagine with no prospect of confirming what is beyond our own universe. Thus, when we speak of a 'beginning of time' we are referring to the beginning of time within our own universe. We call it a universe because it is the only one that we are capable of experiencing. We shall continue to cruise in ignorance of others.

In human terms inertia is lassitude. It is the inclination to continue doing what we are doing in preference to embracing change. Einstein's simple equation $E = mc^2$ is the condensed essence of what constitutes inertia. E symbolizes energy, m denotes mass (the interchangeable equivalence of energy) and C represents the velocity of light. Energy is mass moving in relation to the speed of light. In a three-page paper entitled *Does the Inertia of a Body Depend on its Energy Content?* Einstein argued that energy and inertia each contain a measure of the other. That will slow you down a little!

Constancy vs Inconstancy

It was the tendency for things to continue doing what they are doing that formed the basis of Parmenides' reasoning that change is illogical. What he omitted was the necessary prefix 'all other things being equal', and an acknowledgement that they were not.

We now have to imagine the relation between all things continuing to do what they are doing, a form of order that we have here called inertia, and the competing phenomenal tendency for all things to move in the direction of disorder that we discussed earlier and called entropy. The coexistence of maintenance and decay seem at first blush to be terribly contradictory!

It is this competition that ensures that neither inertia (immutability) or entropy (mutation) prevails absolutely over the other, that the competition will continue and that the constitutional effect, change, is chronic, all serving to explain why the world that we experience is the way that we experience it.

Momentum is inertial existence expressed moment by moment.

38   Tendency: Biased Disposition

Forces force foreclosures. It is the closing of options due to force, or impending force, that narrows the opportunities and inclines events to unfold in particular directions. When we recognize that all matter has properties, and that each property is enabling or denying relations with other entities with other properties, we come to understand that, in considering the relations between any two entities, there are 'preferred', or rather statistically greater, probabilities that they will relate to each other in certain ways rather than in other ways. There may still be competition between forces to determine which prevails over others, but the field of competitors is select. At a certain point in time one force will tend to prevail and dictate the nature of a relationship more than another, at which point we might call the relationship compulsive, compulsion being that state where alternatives give way to single courses of events.

What we have described is a process whereby all possible relations between two bodies are tested and fail the test except one – directed by the resolution of competitive forces. So, if we are human, we might view such a process as a path from the many to the single option. Tendency is the mathematical probability that such a path will be followed.

39   Concurrence: The Seed of Unity

We have recognized the tendency for like forms to congregate in classes and for dissimilar polar opposites to attract, so why do we not all get intimately together in a single clump? Well, we do, universally speaking. How cozy we get depends upon what, where, why, when and how push comes to shove.

Just as forces resolve together to form new forces with new directions, the products of their resolution join with the products of others to do the same. Combinations of causes combine to generate new causes. The act of resolution is an accommodation of differences where the assets of all contribute to the performance of new, more complex units.

The nature of concurrency is the inclination of two or more parties to maintain the same position at the same time, that is, to enjoy coincidence of power. In concurrence we find resolution to act in unison rather than in competition. Operation becomes cooperation. Why?

In the inanimate world all operates according to principles. Gravity says 'come' and everything comes. In this sense, while there is competition for space in the race to the centre of the earth, there is no ill will between its members. All things accept the places allotted them by the forces that prevail. It is only in the animal kingdom, and then only where 'will' insists, that upon occasion we are inclined to violate nature's cardinal dictates.

Incognizance

A clue suggesting the potential for two otherwise opposing forces to act in unison may be seen in the context of a third force that impacts both. When fire ravages a forest, you do not see the lions stalking the deer, the wolves chasing the rabbits or the owl preying upon the field mouse. All the animals of the forest rise to the occasion in the same way because their interests are identical: survival from the peril of fire. Their actions are in accord not because they are magnanimous towards each other but of necessity without concern for appetite or aggression. They may not be cooperating by agreement but they can be recognized as acting in unison. Seeded by adversity, the motivation for their actions is solely self-interest, the agency of its deliverance fear, fear of a fate worse than unity. It is only when dangers recede to the point where behaviour is restored to normal relations that subconscious concurrence may be observed. This is exhibited in the manner in which animals relate to each other as families, packs, herds, flocks, schools, etc. Their primary interest is in individual survival and their secondary interest in maintaining the unity of the group that reinforces their individual security. There is, at this level of development, no comprehension of the interdependence of the species. Agreement in principle, distinguished from instinctive agreement, is an attribute limited to the human species.

Cognizance

In the general scheme of nature, competition between living things occurs between dissimilar species. Certainly trees of the forest compete for nutrients and a place in the sun, availability placing limits upon growth, but they do not wantonly harm each other. Nor do wildebeests or wild beasts of other genre. Each has its own self-interest in survival yet does not maliciously destroy its own kind, with exceptions, and then only under dire circumstances. Indeed, the opposite qualities are more apparent, like the steadfast defence of offspring by parents.

What, we might then ask, are the minimum conditions necessary for mutual acceptance? For inorganic matter and energy, physical conjunction suffices, there being no alternatives; for animals, physical conjunction and circumstantial coincidence of interest brings about mutual acceptance, or at least temporarily abates mutual antagonism; while for humans, physical conjunction, circumstantial coincidence of interest and cerebral complicity may fulfil these conditions.

Simply stated, as an organism evolves in complexity, additional options become available to it by which it may enhance its status vis-à-vis lesser organisms. The human may be as obtuse as a rock, may run with the crowd as an animal, or may use his individual wits to his presumed advantage. The highest degree of sophistication nets the highest degree of benefit. Generally speaking, optimal conditions are shared rather than independently held.

Agreement

When we entertain thoughts of human concurrence we think of reconciliation and agreement. Concurrence presents an opportunity for a less stressful outcome in general than is available through alternative means. Decision Theory tells us that individuals are more likely to benefit if they act in common cause. Qualities that rocks lack, animals exhibit unwittingly but humans employ consciously include intellect, volition and the affections. Potential is revealed through intelligence, intention through judgement, while action is directed with partiality.

Our problem as a species is that we exhibit a mix of individual performance traits that range from those of the rock, through those of the wild animal, to highly intellectual activity. In a democratic society, rock-like and animal-like behaviour constrains us from taking paths to our optimal advantage.

40   Habitat: The Wherewithal

The where, with all necessary for the support of life, is a prerequisite for life itself. Aggregations of matter in differing combinations and subject to differing physical conditions invite interactions that differ by locale. Thus, a particular environment will be more or less conducive to the support of a particular set of relations than another.

Ecologists have modelled the major regions of the earth's biological menages into what they call biomes and have taken inventories of the qualities and quantities of life supported in each.

Examples of distinctively diverse biomes include the evergreen broadleaf rain forests of the tropics, savannah grasslands, deserts, deciduous broadleaf forests of temperate regions, coniferous forests and the tundra. Each is a distinctive ecosystem defined by man to clarify his understanding of the parts of a single system, the habitat of habitats, the earth. The earth is too complex to comprehend as a single idea. At the other end of the scale, each biome harbours enabling conditions for individual plant and animal species to evolve and thrive. What determines what lives where is the match between the ranges afforded by each biome with respect to temperature, moisture and stock of nutrients, and the corresponding capacities of each organism to function within such ranges. All of the above we attribute primarily to the biome and secondarily to the organism.

But there are many animal species that have evolved enhancements to their native capacities to match their environments, amongst the more sophisticated of which may be said to be their skills in modifying their immediate environments in the creation of peculiar habitats. These are the master builders. Master builders exploit their environments in their own interests to enhance their prospects of survival and wellbeing. Such accomplishments have, however, a broader advantage, that of extending the range of the species across the boundary of one biome into another, thus further enhancing probabilities of matching needs with the attributes of the enlarged environments.

Habitats are the domains of inhabitants. Man, in exercising his ingenuity, has developed personal habitats suited to almost all biomes, by which means his habitat extends across all regions of the earth's surface and for short periods in the domains of the subterra, the ocean, the atmosphere and in space beyond.

Mankind protects the interests of resident human inhabitants by conferring regional citizenship upon them. Citizenship conveys priorities in claims to security for citizens. Today, a passport confirming citizenship takes on much the same character as a driving licence, identification not categorically accepted or conveying privilege beyond the bounds of the issuing authority. Trending from man's transience are unions of authority that confer recognition upon people as citizens of larger territories without the necessity to withdraw local privileges from local residences within local jurisdictions. A travelling man may and often does carry more than one citizenship and passport. For mankind biomes are becoming blurred to suggest that we are gradually graduating towards world citizenship.

41   Growth: Change by Degree

Pythagoras taught us that the language of nature is numerical. Through the devices of addition and multiplication we have treated the subject of increase in quantity as a function of numbers. The realignment of matter and its propensities for forming alliances have accounted for variety in composition. All this occurs over time in such a manner that each new condition may be credited as deriving from a simpler form at an earlier time. But change is not unidirectional.

Insofar as all matter is recirculating or in a state of transformation the total quantities of matter do not change but rather transpose their states, affinities and relations.

Accumulation and Attrition

Increases and reductions in any particular form, state or set of relations occur in the course of time by natural composition and disintegration. Such accumulation occurs in essentially the same way regardless of scale. Sediments respond to the forces in a small stream much as they do in a large river. There is no new dynamic that derives from such rearrangement; nothing evolves that has not evolved before to a greater or lesser extent within a similar context. There is no fundamental novelty in such cumulative changes; outcomes are more or less of the same genre.

The changes that we have been describing are the products of basic forces, notably those of gravity, which jostle particles in space, and of the weak nuclear force that establishes affinities between particles based upon their positive and negative polarities. They are inorganic. By this limited definition growth is analogous to increase. However, change works to promote increases 'and' to diminish quantities. Thus, this limited definition is inadequate. A more common understanding is that change brings additions and subtractions; discrete quantities of accumulated things 'grow' larger and smaller. In this sense growth is synonymous with change.

Organic Growth

There are, however, certain combinations of matter that are self-extending, not in the sense that the velocity of a body accelerates due to gravitational forces, distance and time, but rather as a function of native proclivities inherent in certain composite forms. Such combinations 'feed' on the matter in their immediate proximity much as an iron magnet attracts or 'feeds' upon iron in its vicinity, enlarging the contiguous body of iron. In so doing aggregations form simple cells that optimize their capacity to grow to a critical size dependent upon their geometries, much as aggregations of atoms form critical masses that we call molecules. Beyond such critical masses cells are cast off or fail to adhere to existing aggregations and are thus free to form new affinities and new cells. We shall discuss such mechanisms of growth in greater detail in due course; suffice to say such new mechanisms of growth carry unique capacities for the promotion of change and growth that inflect evolution in hitherto untried directions.

The Mathematics of Growth

The events that mathematics describes are natural, but mathematics is not. Mathematics was invented in order to account for observations of the natural world. Because the format of maths proved to be so effective in representing experience in abstract terms, mathematics became the preferred path for the reconciliation and storage of knowledge about the universe. Maths functions like an amplifier of thought whereby dynamic relations may be better understood. A simple equation balances dissimilar relations in a way that serves to clarify or even explain why they are so. Differential calculus provides a means by which the rate of change of variable functions may be represented by changing patterns of numbers that enable them to be rationally compared. The mathematics of growth generates the growth of mathematics. The same mathematical patterns are discernible in the evolution of consciousness and thereby growth in capacity to think and act.

Childhood is a microcosmic representation of that process.

42   Environment: The Roundabout

The environment, all that is in proximity to any thing, comprises a multitude of dynamic influences. One might distinguish them in the classical sense as 'orders' since they all have an ordering function. Let us look at a few of them.

The temporal environment time-sequences all events, permitting and denying future events as determined by which branches of the tree of opportunity are followed and which passed by. One cannot predate one's father, notwithstanding the alluring ditty *I'm my own Grandpa'*.

Then there is the spatial realm that facilitates distancing. Since no two things can share the same point in space at the same time, spatial separation becomes the default option for those that are not where they 'want' to be.

The physical environment encompasses all those causes and effects that we identify with energy and matter in their competition for location in the two environments just noted. This would embrace the physics that tempers the mix at all scales and the chemistry that dictates elementary relations. The physical focuses upon articulating distinctions in kind.

Then we have the economic order that deals with numbers that ultimately determine how many relations can be forged. Economics places limitations on what is possible by dictating where and when resources are limited or exhausted. Thus, economics is the master of opportunities.

At the human level we have the socio-political order, the function of which is to exert non-violent controls over what individuals within societies may and may not do, or should or should not do, and the military environment that exerts controls through recourse to violence when the socio-political system requires reinforcement.

Rather more abstractly, the utilitarian order, Vitruvius' *Commoditas*, equates thoughts and actions with purpose, while the aesthetic environment facilitates spirituality.

None of these latter civilizing orders of our environment are, however, possible in the absence of an intellectual environment, for it is in this realm that idealism and goals are born and nurtured, which form the foundations of progressively higher levels of understanding and worthier goals.

We have of course been referring to order as classified arrangement, as ways of looking at the environment, but must recognize that often-turbulent relations extend between these classes of experience. What goes on between domains as boundary disputes (of which there are in number as many as one's economy of intellect will admit) and how their mutual effects compound to form ever-increasing complexity, are the subjects of succeeding investigation.

The proposition here is that these and other natural parameters, all-be-they defined by man, have been and will continue to be the enduring 'orders' that govern the architecture of all things and their relations, constituting the environment of all-there-is. Environmental orders are relationship facilitators.

The environment appears to be the way we perceive it because of the way we perceive it. If we were to view it within the infrared or ultraviolet wavelength bands rather than as we do, we would have very little to say, for example, about colour.

The premise of Aristotelian and, indeed, all other physics seems to be that all things and events occur in a context. If so, would this not also be true for the universe and, if so, what is the nature of the context in which the universe operates?

43   Boundary: Marginal Territory

Each of us has a clear understanding of the concepts of earth, ocean and atmosphere. It is at the junctures of these that we endeavour to confirm their independent identities.

Confusion

Such a place is at the water's edge. A wave breaks and we sense the realm of the ocean. But as the water withdraws we are conscious of a million little bubbles, bubbles that belong to another realm. The atmosphere has invaded the ocean. We observe that between the bubbles the water is not as we should reasonably expect, clear and transparent, but clouded and turbulent, having the behaviour of water but the appearance of sand. The earth has taken leave of its sedimentary disposition and been whipped into a frenzied dance by the action of the water upon it. The certainties of what and where are less so. The notion of ocean is confused. As the waters depart we become aware that the sand is not solid at all but perforated by an infinite number of little holes forced open by the discharge of air. The winds lift, blowing the spindrift off the tops of the waves. Along the beach we observe miniature tornadoes of sand eddying towards us until they pepper us from head to toe.

We are, we know, in a breathable environment yet we cannot deny the waters of the ocean that spray upon us, or the sandpaper textures that our skins have assumed. Another wave breaks but this time we are a little less confident of where we are. Our understanding has become clouded. We become boundary markers between the three great domains at this particular place and time. We have lost a grain of presumptuous knowledge and gained a grain of wisdom.

So it is when we examine carefully the boundary conditions of existence.

Fusion

While boundaries in nature are often diffuse, approximate and changeable demarcation lines that differentiate one composite form from another, for us they are fundamental to our perception.

All entities have boundary conditions that describe their extents. For the living it is our membranes, for the inorganic it is distinctions in matters of fact. Across these boundaries percolate small quantities of immigrant and emigrant matter and energy, small, that is, compared with the pools from which they derive, but often large in their significance to the entities to which they relate. Thus, boundaries are approximations of juncture, which fact is often evident when examined by means and at scales other than those of our natural perception. Examination of the microscopically small tells of almost infinitely extended boundary surfaces, fractal in character, being composed of oft-repeated patterns at ever-diminishing scales, while focus upon the macrocosm reveals the tenuous nature of boundaries that we otherwise assume and would like to read as sharp lines of demarcation.

The boundary conditions between one thing and another are not as simple as we often may think, as the idea of the fractal reminds us, but we are unable to conceive of a totality as complex as the universe without mentally dividing it into smaller packages. The boundary lines are drawn where minimal exchange takes place between one zone and the next, to bolster our abilities to conceive of and communicate ideas as much as to define entities in absolute terms. By such means we distinguish a tree from trees, and trees from a forest, each characterized as a critical mass.

The most efficient boundary is that which encompasses the greatest area or volume with minimal extent. In the world of two dimensions the circle performs such a minimalist function and the circumference is represented by the expression $C = 2\pi r$; in three dimensions it is the sphere, the surface area of which is described by the expression $A = 4\pi r^2$. In each case these forms do not nest together without creating interstitial spaces between, zones of contention as to what they comprise and what forces govern.

44   Form: The Expression of Relations

Having stated the state of state earlier as arising from the subjugation of something to external conditions, like temperature, we need to recognize that there are other forces that give rise to the formation of matter in aggregate masses that are not environmental in nature.

Form as Structure

The first of these we would acknowledge to be self-directed, that which results from a thing holding itself together intrinsically as a consequence of its particular and essential nature, which quality we might call 'form as structure'.

Form as Shape

Then there is the classification of form as image that we bring to bear upon all experienced forms as a means by which we may distinguish one pattern from another and communicate such distinctions. We idealize such forms, labelling them according to differences in character, as minerals conform to crystalline character based upon the number and relation of their surfaces. This characterization, our perception of categories of arrangement, we might call 'form as shape'.

Each such form is a manifestation of a thing or action, being or doing what it is or does according to its fitness for that purpose.

Missin' Formation

When we experience misadventure it is generally because our relations do not correspond well to the expression of the form that we are in close proximity to. Those for whom it is their misfortune to drown in water do not know that they were surrounded by vast quantities of that essential oxygen which they craved, only to be denied by the inaccessible form that it assumed. A fish out of water experiences the opposite plight to the same effect; it drowns in air. In the architecture of existence form is all-important because it facilitates or denies opportunities for interaction. The classic problem of the square peg misfitting the round hole comes to mind.

Formation

The critical mix of constituent parts and their proximities may, thus, be viewed as, and called, 'form as the expression of relations'. All forms are statements of spatial or temporal order, space and time being the inescapable governing parameters. We make explicit statements about form applicable to a given place at a given time, even if we are uncertain as to where or when they occur. Form is now, in the present, and was likely different in the past and will exhibit further differences in the future.

The architecture of anything and everything describes the nature of relations frozen in time, to each of which we give a name that conveys a form that represents its relations. The distinction between frozen state and frozen form is semantic. As used here the term 'state' describes the focus of attention as being the extrinsic condition or context bearing upon an object, whereas 'form' describes the intrinsic or attributed character of that object. We apply these terms interchangeably consistent with established usage in language. Without such frequent, informal and dextrous reordering of ideas we might otherwise be led to a form (or would it be state) of dementia.

45   Motion: Change in Physical Relations

If there was but one thing in existence, being singular in number and therefore irreferable to another thing, it would be impossible to establish whether it was moving or not. There would be no relations. Similarly, if there were two things at a constant distance apart and with unchanging exposures to each other, it would be impossible to establish movement, they would be as one. In each of these cases a detached observer would be obliged to conclude that there were no discernible changes in physical relations. It is only when we detect a change in the position or attitude of one body as it relates to another that we conclude that motion is involved. Even then, if there were two things 'appearing' to orbit about each other with changing exposures, it would be impossible, as Newton discovered, to establish whether one was moving around the other, or if the other was moving about the one, whether one or both were moving, or that the observer was in motion.

We attribute change to motion, a term that we use to signal the fact of inconstant relations. The rates of change we call acceleration or deceleration. We refer back to our prior examination of forces to account for, or attribute, the de facto origins of motion.

Of Course

The effects of force can be objectively studied through cosmological observation, from which, over time, patterns of relational behaviour of celestial bodies may be confirmed and elaborated.

Insofar as we observe movement through our eyes, the objects, directions and extent of movement are always discernible relative to the location of the observer. Thus, it was not only natural but logical for Aristotle to believe, with his feet planted securely upon the earth, that the earth was stationary and that all the objects visible in the night sky were orbiting around the earth. In the second century A.D. Claudius Ptolemy organized the knowledge of his time into a relational model intended to account for the differing patterns of movement observed in different bodies. On the understanding that some bodies emitted light while others merely reflected it, Nicholas Copernicus structured a cosmological model on the premise that reflectors orbited larger emitters and that the planets moving along circular courses orbiting the sun were exemplary.

With the invention of the telescope it became possible to observe the planets in greater detail. In viewing Jupiter, Galileo observed the planet to have several lesser bodies orbiting it much as the moon orbits the earth, suggestive of a hierarchy of relations based upon mass and proximity. By experiment Galileo was able to show that falling bodies increase their speeds at the same rates independent of their respective weights. What this meant to astronomy was that the effects of gravitational forces were predictable, constant and calculable, thus eliminating one layer of uncertainty with respect to the behaviour of celestial bodies.

Off Course

Concerned that circular orbits did not accurately account for the cycling periods of the planets, Johannes Kepler envisioned a model based upon elliptical orbits. If you rotate a circle, say a ring, 360 degrees off its original plane of alignment, at only two points in its course, at 90 and 270 degrees to its plane, will you see a circle. From two viewpoints, 0 and 180 degrees, you will see a straight line, while from all other angles you will see ellipses. The converse is also true; that is, an ellipse seen from a particular viewpoint will have the appearance of a circle. Aristotle, Ptolemy, Copernicus and Galileo had accepted and trusted their understandings that circular appearances required the presence of a circle. Kepler opened up an infinite range of possible views of a circle, by degrees and subdivisions of degrees. By this means the longer axis of an ellipse could be adjusted to 'fit' and account for observations in relation to the shorter axis. Cosmological theory and observation appeared to be reconciled for the first time.

The drag of inertia visible in the delayed tidal actions of the oceans and the distortion of the earth's spherical shape due to centrifugal force (accounting for the greater diameter of the earth at the equator than around the poles) mean that the sphere is in fact a near-sphere, and its field of influence, including its centre of gravity, could vary subtly. Such variation could contribute to the distortion of other circular orbits into ellipses, as could the changing relations between neighbouring bodies variably effecting the strengths of gravitational forces between them. The ellipse lends itself to changes in axial proportions where the circle does not. The 24-hour day has not always been so, nor has the 365-day year; they are very subtly changeable. The ellipse responds to those forces of change where the circle cannot.

On Course

Motion in its simplest form has been described by Newton as that which is ongoing in straight lines as determined by inertia. His first law of motion reiterates Galileo's finding that, in the absence of interference, continuity, whether of rest or movement at constant velocities and in constant directions, is the natural course of behaviour of all matter. Could this underlie the profound conservatism and desire for immortality in man?

All other forms of motion Newton held to be complex to the extent that they are energized by forces acting on a body in motion in addition to inertial forces. What happens when such changes occur is the subject of his second law, which reconciles disparate forces in their quest for resolution.

The Course

When a force that acts upon a point in one direction is opposed by a second force acting in the opposite direction, the greater force will offset the lesser force and the remaining balance of the greater force will be applied in the direction of the greater force. This is the principle exhibited in the 'tug-of-war' (pulling in opposite directions), or in 'arm-wrestling' (pushing in opposite directions). When two forces act in the same direction upon a point, their values are combined and act in that same direction. This is the principle exhibited by two locomotives pulling or pushing in tandem.

When two forces converge upon a point at any angle other than 0 or 180 degrees, the value and direction of the resultant force may be represented by the diagonal of a parallelogram the sides of which are proportional in length to the values of the forces, and parallel to the directions that the forces act. By way of example, a point acted upon by two forces of values 3 and 4 at 90 degrees to each other will translate into a resultant force of value 5 in the direction of the diagonal of a parallelogram, in this case a rectangle, of proportions 3 to 4. Similarly, a right-angled triangle, with sides in the proportions 5 to 12, representing values and directions of forces acting on a point, will resolve in the direction of the diagonal of a parallelogram of those proportions with a measurable value of 13. From the two examples above we can conclude that the greater the disproportion in the original forces, the closer the value and direction of the resultant will be to those of the greater force. Pythagoras would be delighted to learn that his findings concerning his ubiquitous 3:4:5 right-angled triangle underlie the most elementary fundaments of mechanics.

Physics may be explained in terms of geometry, or geometry in terms of physics. Newton's inverse square law of gravity is the physical equivalence of Pythagoras's geometric square of the hypotenuse theorem. Each relies upon two variables to define a third, Newton transposing Pythagoras's two-dimensional format into three dimensions through the introduction of mass.

The parallelogram reconciles forces acting in straight lines much as the ellipse reconciles forces in orbital motion.

For Cause

Force is the causal factor governing displacement or change, while mass and distance govern the rates of change.

Taking Newton's discovery that the physical relation between two (inert) bodies with respect to mutual influence (force) is a function of their masses multiplied by the square of their distance apart, it can be shown that where no separation exists, proximity substitutes. The proximate boundaries of each body, being in contact, become effectively one and the same, describing a new common boundary condition. In this mode the effects of the forces are maximized and can act and counteract with the minimum of loss due to transmission. In so doing, they can carry the essential qualities of each to the other. In the absence of distance the forces 'are' the relations with respect to motion. Newton's third law of motion stating that for every force there is an equal and opposite force is not simply a pragmatic picture for the benefit of science, it is a poetic expression of understanding of the relations between anything and anything else with respect to mutual influence.

46 Velocity: The Rate of Change

Having discussed mass, inertia, force and motion, we would be remiss to continue without a few words concerning velocity.

In the Abstract

We have already identified a standard by which we might gauge the velocity of any particular body in motion as a ratio of an abstract maximum velocity possible; that enjoyed by light. If it were not for the practical difficulties of measuring such ratios we would no doubt treat all velocities as factors of 1 or percentages of that theoretical maximum; this would be the most logical comparative method. By the same token logic would suggest that we measure heat from the base temperature of absolute zero (−273.16°C or −459.69°F) rather than on an arbitrary basis that Lord Kelvin or Gabriel D. Fahrenheit conceived, but this is not practical, and we do not. We therefore seek a measure of practicality by adopting an accessible base line and convenient units for measurement.

In Relation

The unitary measurement of velocity requires two frames of reference: time and distance. Man has adopted satisfactory measures for the passage of time that have been universally endorsed. Standard measures of distance have evolved relative to special circumstances. Such standards are readily comparable and interchangeable for convenience.

All but the most abstruse of animals have discovered that the best way to get along in life is to move along the alignment of their major axes. Even a baby knows this. If one measures their performance in feet per second, their feet, some will be seen to perform extraordinary feats. But some do not have feet! Comparisons between designs by nature and those by man intended to overcome certain forces of nature, when measured in body lengths per second when moving at near top speeds, reveal telling implications concerning relative efficiencies of their designs and the resources required to propel them a distance of one body-length.

Compare the following in descending order of performance measuring velocities as a function of distances moved in body-lengths per second: bees − 880; swallows − 270; cheetahs − 30; production automobiles − 8; man − 5 (eccentrically moving along his minor axis but measured relative to his major axis); commercial jet airliners − 4; and the shark − 1.

Note that the swifter shifters are moving in still air and have small masses. As they gain in mass or resistance due to contact with the ground, or their paths become warped by gravity, or they move in a denser medium like water, their relative velocities in body lengths per second diminish drastically.

In Conclusion

In order for a worldly body to gain the highest possible velocity it would require being in gravitational free fall, have the massless quality of a photon and to freely pass with minimal resistance through the atmosphere. That is just what light is, has and does on is way to the earth.

Since we use neither of the above-noted logically pure systems of measuring velocity, the single standard (relative to the speed of light) or the subjective standard that denies comparison (body-lengths per second), we have acquired arbitrary processes of measuring velocity based upon linear systems in local usage, as miles and kilometres, that constantly require interpolation into other systems for the purposes of achieving comparisons: a kind of hybrid comparative systems comparative system! At the close of the twentieth century the world was awash with such illogic pertaining to measurement systems of all kinds.

47 Evolution: The Chronology of Change

Forces exist that cause effects. The effects take many forms determined by the contribution of the forces, and of the environments that they operate in that facilitate or deny certain events to occur. Forces force change. Time permits these events to occur in succession such that, in hindsight, they can be read as histories, each particle describing a unique path in space and time. It is this unfolding historical development of successive changes that we broadly describe as evolution, using the same term to more narrowly describe the same processes as they relate to organic changes in the histories of the living.

Evolution in mathematics is the process by which root numbers are discoverable, a process analogous to looking backwards into the ancestral lineage of a number to show how, in time, novelty and complexity issue as the products of simpler numerical relations.

Controversy over First Cause

Recalling our earlier discussion wherein we noted our seeming compulsion to establish first cause that we have traditionally attributed to a deity in the absence of plausible evidence, we have but two possible explanations. The first holds that all began, either literally or figuratively, much as the book of Genesis describes, with creation as a fantastic spontaneous metaphysical event whereby nothing suddenly became something, the cause of which God only knows. This has its parallel in physics with the Big Bang theory where all devolves from the infinitely small. In each case there would be an act of creation, and a beginning of time and everything else. The second explanation would be that all the elementary parts that exist today have always existed, that change is a constant condition, and that through change creation is confined to reconfiguration, the cause of each change being a realignment of forces. In this second scenario there is no cause to attribute first cause, and no point to attributing a point of beginning to time.

While each theory would appear to exclude the other, there is room for reconciliation if we accept the biblical explanation as non-technical, the physics as solely pertaining to this (our) universe, and the infinite duration occurring in the context of an infinity of space that would include any other phenomena (gods or universes) of which we have no knowledge. In the absence of proof we are left with the luxury of choice to believe whichever explanation is more consistent with our experience or mental comfort, recognizing that the future realignment of forces as causes may have the effect of changing our minds.

By this resolution we need no longer feel that evolutionism and creationism are contradictory; they merely represent points of view that place more or less stress upon a need to look further for explanations. They both acknowledge the experience of sequenced change over time. A natural and seldom-stated objection to evolutionism by so-called creationists is that the theory of evolution, by implication, depicts those with fundamentalist metaphysical beliefs in creation as being more primitive, evolutionarily speaking, than those that seek and find rational explanations for change in the processes of evolution.

Urge to Account and Hold Responsible

Our next obsession with cause addresses precedence of authorship in order that we may establish responsibility for subsequent developments, praise or blame according to our view of each event as being beneficial or detrimental to our interests, and thereby show cause to reward or punish in order to justify our simian impulses to control our immediate environments. In our actions we attempt to command evolution; in our reactions we assign blame and punish ourselves for our failures.

Propagation of Change

Forces occur in time in serial to, parallel with, divergent from or convergent towards each other.

Thus, it can be understood that history lines of multiple particles will do the same, carried along by the stream of events. Serial or parallel courses suggest inertial tendencies to continue moving in straight lines, while convergence and divergence imply collision paths that trigger changes in course. It is not difficult to recognize that evolution is influenced similarly. Novelty can only occur when there is change, i.e. of direction (quality) or number (quantity), as when forces or particles divide or combine. Where there is division there is subtraction in quantity from the source(s). Where there is combination or multiplication there is addition in quantity to the product. Division is identifiable with subtraction as negative, while multiplication is identifiable with addition as being positive, as yin is to yang. This extends across the threshold between the non-living and the living, the only difference being the nature of cause, being external for the lifeless and internal for the self-reproductive; a tenuous distinction.

The Darwin/Wallace Convolution

We will take a moment to remind ourselves of the coincidences and ironies that attended the lives of two nineteenth-century English naturalists, Charles Darwin and Alfred Russel Wallace.

Born within 70 miles of each other in the England – Wales border country, college educated with no particular stress on the subjects to which their interests were later to gravitate, each was closely associated with land surveying activities, which kindled their interests in geology. In their early twenties they made independent expeditions to the equatorial region of the east coast of South America, where their interests in botany and zoology were aroused.

Darwin continued on a five-year voyage around the world in the survey ship HMS Beagle, a voyage best remembered for the extraordinary number and variety of plant and animal specimens shipped back to England, and for his observations of thirteen distinctly different species of finch unique to the remote Galapagos Archipelago, which distinctions catalyzed his theories on evolution and formed the foundation of his later publication on the origin of species.

Wallace's return to England is remembered for the loss of his collection of specimens to fire aboard his ship and the publication of his *Narrative of Travels on the Amazon and Rio Negro*. Four years later he was to embark upon an eight-year tour of the Malay Archipelago, where he observed a sharp demarcation to the east of Borneo and Bali between classes of animal life that had affinities with those of continental Asia, and those that were identifiably Australian in character, a division known today as the Wallace Line. Wallace originated his theory of natural selection at this time, incorporating its essential premise in his essay *On the Law which has Regulated the Introduction of New Species*.

Both Darwin and Wallace had been forcefully influenced by Thomas R. Malthus's *Essay on Population* in which Malthus postulated that populations always increase faster than their means of subsistence. Darwin was inspired to recognize that favourable variations would tend to favour preservation, while Wallace was prompted to conceive his theory of the survival of the fittest. What was notable for its absence at that time was any evidence in support of heredity, which fact placed a heavy burden upon any proponent of a theory of evolution. The burden was removed with Gregor Johann Mendel's theory of units of heredity (genes) as particular carriers of memory that, due to their exponential potential to form linkages in chromosomes, are prone to produce new variants by mutation, developments that ran concurrently with the work of Darwin and Wallace.

In 1858 Wallace mailed the manuscript of an essay to Darwin (known to Wallace by reputation as a former Secretary of the Geographical Society and his senior by fourteen years) the contents of which revealed, according to Darwin, a startling and complete synopsis of his own theory of natural selection. In Darwin's words, 'If Wallace had my manuscript sketch written out in 1842, he could not have made a better short abstract.' Darwin referred the matter to his peers and it was decided that Darwin should make a presentation to the Linnean (Zoological) Society of an abstract of his own work and of the Wallace manuscript entitled *On the Tendency of Varieties to Depart Indefinitely from the Original Type.*

Darwin was placed in a difficult position. On the one hand he had in hand irreproachable corroboration of his own theories, on the other hand he had in Wallace a serious contender for the laurels that would eventually fall upon the one who articulated a theory of evolution first and with the greatest authority.

It is not known whether Darwin was acquainted with the two Wallace publications having a bearing upon the subject of their mutual interest prior to Wallace's direct communication to Darwin, or whether Wallace had read any of Darwin's publications prior to the Wallace letter to Darwin. Darwin had the distinct advantages of arranging his own abstract after receipt of Wallace's paper, and of being the recipient of a government grant that enabled him to devote the balance of his life to research and writing, and the disadvantages that his 1842 manuscript sketch *The Foundations of the Origin of Species* had not been published, and the exposure to possible criticism for representing that Wallace's terms 'now stand as heads of my chapters'. Why did Darwin wait until 1856 to commence his treatise eventually to be known by the title *On the Origin of Species by Means of Natural Selection, or the Preservation of Favoured Races in the Struggle for Life?*

Ultimately, that necessary authority was shown to be vested in Darwin, his salient contribution being to show 'how' evolution worked, triggered by the seeds that lay, waiting to be broadcast, in his descriptions of variety in the forms of the beaks of the Galapagos finches, which he attributed to sparsity of and disparity in food types in remote and separated island environments – buried in his first publication, *Journal of Researches into the Geology and Natural History of the Various Countries Visited by HMS Beagle*, 1839, later released as *The Voyage of the Beagle.*

Parallel and Convergent Evolution

What we have described is exemplary of parallel developments, in this case of human thought which, through Wallace's communication, converged, prompting Darwin to accelerate the completion and publication of his *Origin of Species*. If Darwin's shipload of specimens from South America had burned as did Wallace's, and had Wallace's survived, how would their respective contributions and reputations compare today?

The coincidental parallel development of ideas, as was the case for Darwin and Wallace, is not merely unusual, as Newton and Leibniz experienced to their mutual discomfort upon discovering that each had independently developed a new branch of mathematics later called calculus, but perfectly normal. It is normal for like kinds to dwell in like environments, to have like experiences, and to evolve simultaneously and in similar directions as determined by their shared assets and environs. What is uniquely human and recent about this phenomenon is the advent of mutual awareness. This is evident in the growing tendency, for example, for Nobel prizes to be shared by progressively larger numbers of ostensibly independent recipients.

What appears to be necessary for common development is inherent likeness, of which common inheritance and common experience are the common denominators.

A revolution is a full cyclical passage of one of nature's processes. We misinform ourselves as to what the word means by inferring that revolution is what happens when we attempt to abruptly break away from such a pattern of events. In so doing we transpose the original meaning to its opposite meaning. This second meaning is simply giving emphasis to aspects of a new cycle that were not so pronounced in prior cycles. Evolution is the historical accumulation of revolutionary experience.

Destiny of Species

We return to our working definition of evolution that reduces to the tendency for complexity and variety to increase as a mathematical function of all systems. In this regard evolution may be equated with entropy, but in general we tend to use the term entropy to describe the compounding complexity of relations between inanimate things, and to reserve the use of the term evolution for discussion of the tendency towards exponential development in living things. This inconsistency is a function of common usage rather than due to any literal distinctions.

Evolution appears to be harnessed to two postulates, that entropy will continue to ensure the flux in elementary particles stirred by forces endeavouring to compensate for the unequal distribution of matter in the universe, and that new relations attributable to such flux, particularly in genetic materials and social relations, will invariably produce individual organisms that are better able to survive and thrive than others, and that will thereby tend to propagate and dominate.

Succeeding discussions suggest what the human species may be evolving towards.

Sequence and Consequence

A way of looking at evolution is suggested by viewing the development of an individual life from inception through old age as a decision tree that maps alternative possibilities that are selected or denied by circumstances. Such causes and effects, multiplied innumerable times, collectively confirm Wallace's conclusions regarding *the Tendency of Varieties to Depart Indefinitely from the Original Type*.

48   Earth: The Habitat of Habitats

Gaea

Earth as a system of systems is what it is because of the puff, slosh and grind rates of interaction between the earth's atmosphere, oceans and soil, tempered by life and aided and abetted by the intra- and extraterrestrial. The global puff rate is just one year, anywhere to anywhere else by air mail, the slosh rate is 1,000 years from here to there or a round trip halfway, while the grind rate is measured in geotectonic time per cycle, at one inch per 1,000 years.

Fuzz is where the buzz is. Fuzz, that ubiquitous yet equivocal interzone between the terra firma of the planet and the clear blue, is where all meets all else. It is where the meteoric thunderbolts from outer space and the volcanic thunderbolts from the inner core are arrested in their fiery courses. It is where the radiant heat of the fire sun splashes to a halt. It is the grilled ham sandwich from the brown bread-coloured organic soils to the white bread-coloured fleecy atmosphere. Keep your eyes on the ham for it may not be what or where you think it is.

All these manoeuvres of matter, this continuous organic/inorganic dance, engineer unions by which each benefits from the other's attributes – all propelled by energy.

The earth is analogous to an automated science experiment whereby an almost infinite number of relations are generated between myriads of entities, many of which forge affinities, dependencies and symbiotic unions, not by design, but as the products of mathematical probabilities and physical and chemical activity. Those experiments that 'work', i.e. that serve to support the recurrence of more 'work', tend to modify other relations that are otherwise inert. It is the 'working' that accounts for evolutionary processes, and the compounding of 'successes' that yield more and more complex relations.

Life and consciousness were very low probability outcomes of such varied permutations, made more probable and eventually possible in time through a near-infinite number and variety of mixing and matching recycles. Our system of systems is a numerical permutator, to no purpose but with wondrous and diverse effects. Eventually, perhaps inevitably, it was to deliver conditions whereby 'will' and 'consciousness' emerged with the potential and discipline with which to begin to comprehend the processes through which they themselves had emerged yet remained captive. The prospect of enhancing control of the earth's magnificent dynamics to further favour the living over the non-living, a kind of 'Gaean' self-governing culture headed by human intelligence, is a long shot from our strenuous aspirations to promote forms of democratic government that dilute intelligence with intellectual poverty. If ants conducted opinion polls, they would be all over the place. Well, so they are! Does that mean they do, and as a result have risen to their highest level of competence?

Future events remain concealed with regard to whether the captor or the captive may serve to safeguard the 'who' or 'what' from untimely extinction. The earth 'owns' itself. We are all transients upon her, like so many ticks, fleas or lice on the hide of a beast, rendering symbiotic services to, freeloading upon or inflicting damage to the system as the case may be, during our short tenures upon her.

Who can say with authority that the earth, indeed the cosmos, is not as cosmetic powder on the face of a larger reality?

49 Atmosphere: The Heavy-Duty Lightweight

Zeus

The spinning planet puts spin on its atmospheric veil. The spin eddies at regional and local scales subject to attendant modifiers, the greatest of which is temperature. Thus, north and south of the equator we have opposed systems of air circulation driven laterally by the earth's spin and vertically by temperature. It has been estimated that a full quarter of the solar energy admitted into the earth's cocoon is applied to changing the state of water to water vapour, those fluffy little sheep that slither across the blue nothing from who-knows-where to who-could-ever-know-where! The hot air, sated with moisture, rises and cools. Unable to retain its moisture in the cooler environment, condensation releases rain. The two opposed systems, known as the Hadley cells, wrung dry and mutually inhibited from crossing their common boundary, spiral away from the equator and descend as dry air at about latitudes 30 north and south, accounting for the major deserts in those regions. Without these worldwide whirls, our regional extremes of temperature would be more so.

Eden

Surface geography has modified the effects of this *perpetuum mobile*, making local desert areas more hospitable to life. Nowhere was this more evident than at the ancient site of Babylon, that, while it may have enjoyed a temperate climate following the recession of the pleistocene (glacial) epoch more than 10,000 years ago, owes its legendary reputation as the locale of the Garden of Eden and site of the 'hanging gardens' to its strategic position on the fertile alluvial plains between the Tigris and Euphrates rivers.

When we investigate this atmospheric turmoil we ask ourselves what is the matter and discover that the matter comprises about 78 per cent nitrogen and 21 per cent oxygen. The remaining 1 per cent includes all other atmospheric gases: the so-called greenhouse gases, water vapour, carbon dioxide, methane and nitrous oxide, and traces of many others including ozone, helium and hydrogen. Why should we be surprised that the elements helium and hydrogen are so heaven-bent on escaping from the earth's influence? They are just doing what comes naturally to all matter, seeking relief in an incongruous environment: they 'want to go home' where they comprise about 25 and 75 per cent of the mass of the cosmos respectively.

How this apportionment of atmospheric gases came to be as it is, is a subject inseparable from the fact of life itself. At 21 per cent of air, oxygen reaches peak production at the end of the summer when broadleafed plants over-compensate for the emission of carbon dioxide. In the late spring the reverse condition occurs when we have a small deficit of oxygen. But in the opposite hemisphere the cycles are reversed, which tends towards the maintenance of global stability in oxygen production.

The Wizardry of Oz

Ozone occurs naturally in the stratosphere as the product of the barrage of low frequency ultraviolet light upon oxygen, fusing its atoms into triple-bonded molecules. While ozone derives from oxygen that derives from life, its effect is to absorb short-wave radiation that would otherwise parch the earth's surface, so affording life on earth a blanket of protection from the scourging effects of an overdose of generosity from our patron star the sun.

The effect of the atmospheric wrap around earth is to allow visible light to pass inbound but to block the escape of a large proportion of the reflected heat, infrared radiation. This trapped energy is either reflected back to earth or absorbed by greenhouse gases like carbon dioxide, but primarily water vapour, which precipitates after about a one-week flight, carrying its payload of energy back to earth. In doing so the average surface temperature is raised by about 60°F (33°C) above what it would otherwise be in the absence of atmosphere. A natural balance is maintained by an approximate equality in the supply of, and demand for, carbon dioxide by living organisms.

Mr Hyde

The maverick in this natural greenhouse gas equation is man, who contributes industrial emissions to the atmosphere, including excess carbon dioxide, while not providing a compensating demand for these gases. Carbon dioxide files a 100-year flight plan before returning to the surface of earth to be reabsorbed into the oceans and plant life. As a result, excesses compound to increase the rate of global warming. While the average surface temperature has risen in the order of magnitude of one degree in 100 years, we should not look at numbers but rather the effects upon us that statistics like that represent. That particular number has accounted for the meltdown of glaciers and otherwise permanent ice fields, which have contributed to a rise in the mean sea level of almost one foot. Ironically, some of our emissions, and those of volcanoes, including sulphates, serve to block incoming radiation in the short term.

Dr Jekyll

Our ability to alter the climate is incontrovertible. What we do or fail to do in the future in this regard is potentially the subject of one of the greatest controversies of the civilized era.

What underlies this surcharge of the atmosphere is concern as to how it will affect mankind and life in general. Pluses and minuses don't add up if there are certain uncertainties. With this interest in mind science addresses possible outcomes through mathematical modelling.

Perhaps only human intervention in human intervention can forestall succession of the worst-of-all-worlds that may otherwise evolve by default.

50   Ocean: The Earth's Exotic Potion

Poseidon

The ocean is, literally, the pool of all. All elements can be found here in their ionic forms. The significance of the ion (found in salts, acids and base materials, often in solution) lies in the fact of its having additional or subtracted numbers of electrons from those in a typical atom or molecule, which event reverses their electrical polarity. By this means, positively charged atoms become negatively charged and are attracted to other charged atoms from which they would otherwise be repelled. In consequence, exchanges and unions, including life itself, occur that would not in simpler states. It should not come as a surprise that the most abundant ions in human plasma are the same two, chloride and sodium, that prevail in the ocean, yet the biomass in the ocean constitutes a mere 1 per cent of that of the planet earth as a whole.

The ocean is the primary pool for the element carbon and estimated to contain all but one part in five of all the sulphur of the planet. Notwithstanding that, sulphur plays a critical role in life itself and is found in certain amino acids, the progenitors of proteins.

The oceans are in a constant state of restlessness. The causes are numerous and their interactions confounding. Let us examine a few of these.

The Push-Me-Pull-You

The earth's gravity demands that all move downward towards the earth's centre. The moon's gravitational forces insist upon pulling all matter towards the moon with forces proportionate to the bodily masses involved and their distances apart. This causes deformation in the earth's forms, most conspicuously in the more massive, fluid and contiguous bodies: the oceans. This drag, that we refer to as the action of tides, particularly as they relate to confined passages where we witness ebb and flood, is confused by the constantly changing position and distance of the moon relative to any particular local body of water.

Round and Round

The earth rotates about its axis relative to the sun (counter-clockwise as viewed from the North Pole). The speed of surface displacement is greatest at the equator, where all points are at maximal distances from the axis.

The oceans, mediated by inertial forces, rotate eastward at slower rates than the land masses to which they relate. As a consequence, when viewed from an adjacent point on land, the ocean waters 'appear' to move westward along the equator.

As with Hadley cells of the atmosphere that spiral vertically and away from the equator, these ocean currents wind away from each other, clockwise in the northern hemisphere, and counter clockwise in the southern hemisphere when viewed from above, sloshing back in the opposite direction closer to the poles where the earth's rotational forces are weaker. Since the impetus behind the westward movement of warm surface water at the equator (assisted by the westward-blowing trade winds and suffering the same effects of inertial drag) is much greater than the relatively slower forces of flow on the return cycles, there is a tendency for water to 'pile up' in the western Pacific. Since water doesn't pile up very willingly, it does the next best thing and backs up. Over a few years this can cause a 10,000-mile traffic jam from the Philippines to Ecuador. A corresponding cycle of atmospheric change, the Southern Oscillation, accompanies this event whereby remission in the strength of the trade winds works to aid this back-flow process. Not having anywhere further east to back up, it will deflect north and south along the west coasts of the Americas, flooding low land and raising the surface temperatures of the ocean by 5 – 10°C (9 – 18°F).

Feast and Famine

The plankton count dives in warmer water, and the oceanic food locker becomes depleted for all shapes and sizes up the food chain, for marine bird life and for human fishermen. On the credit side, the warmer air can hold more moisture and this translates to an increase in rainfall over land.

Vegetation thrives, land-based foragers thrive and carnivores thrive. This phenomenon is known as El Niño, 'The Christ Child', originally so named by the fishermen of South America due to the time of year, around Christmas, when warm currents washed their coasts.

Since no increase in solar radiation occurs, where does the increased rainfall come from? The eastern Pacific coast's gain is the western Pacific coast's loss; Australia and Indonesia pay the price with corresponding droughts on El Niño years!

Up and Down

Convection ordains that warm water shall rise vertically to displace cooler water, causing vertical circulation. This is particularly evident where winds blow warm water offshore. Replacement water has to come from lower depths that are cooler, modifying the atmospheric temperatures.

Add to these causes the effects of winds generally, differing water densities, eddies due to the convergence and divergence of currents, deep-water back-flows, interference by land masses and the mutual effects of each upon all others and we have an engine of indescribable productivity. Its product: an enormous subscription of energy applied to the cause of the socialization of matter.

51 Soil: A Matrix of Matter

Hades

Soil, the underworld or 'infernal regions', is the boundary condition between the concretion of the planet and the nebulous air, a complex frictional interlayer that owes its existence to the collective scouring of the spheres atmo, hydro, geo and bio. Soil is thus an amalgam, the composition of which varies to such a degree of diversity as to enhance the possibility that a critical mass of chemicals, so necessary for the formation and maintenance of life, occurs somewhere, sometime. Insofar as this is so, soil has evolved as a biological construct and deconstruct that tends to sustain itself in approximate balance.

The contraction of clay soils due to heat and the evaporation of moisture cause cleavages in the soil. The counter-action of cooling and wetting closes them. Anyone who has tried to cultivate such a soil knows this well. Wet clay becomes impermeable to additional moisture, the expansion and contraction functioning like a thermostat to determine 'go' or 'no go' conditions for the added water, all governed by heat. The same principle applies in different ways to many inorganic materials. Mountains break up or down, due in large measure to the alternation of freezing and thawing cycles that contract and expand their parts. Energy is the 'activator', matter the 'activatrix'.

This behaviour, expansion and contraction, opens and closes the doors of possibilities, not the least of which is to life itself. Viewed in the context of geological time, our summer/winter cycles look like breathing, the surge of tides like heartbeats, habituated by repetition until it becomes uncertain which came first, the muscular spasm or the flow of current, the chicken or the egg, by which time they have become inseparable and interdependent.

The sciences endeavour to map all such actions, reactions and reciprocal relations.

Soil is the home turf of nearly 70 per cent of all phosphorus. What is most significant, for us, is not the distribution of phosphorus by quantities but rather the fact that life is dependent upon small quantities, bearing it in DNA (deoxyribonucleic acid, the basic building blocks of life-bearing systems) and in energy exchange molecules. Phosphorus is not to be found in useful quantities in the atmosphere.

To understand better what is going on under our feet requires an enquiry into the nature of life itself, and of the reciprocal services that the soils and life afford each other.

52 Organization: Vitality Through Interdependence

We have ventured through the topics of existence, number, relations, change, forces, causes, effects, diversity and novelty to discover tendencies towards special relations.

We defined order as discernible pattern. In doing so we accepted that order was of our own making. Order is the simplest form of arrangement, a fixed state of relations that has no greater significance than that of uniting parts into sets that do nothing as a group beyond triggering such concepts of combined relations in our minds. The parts of patterns may change in their relations but as long as they continue to signify no more to us than rearrangements of pattern, they remain simply orders undergoing reordering.

Mechanisms

But when we discern a process the effects of which rely upon cooperative parts working in unison, we accept the concept of reordering as fundamentally different and confer upon this combination or process the term 'mechanism'. What we are recognizing is a structure of interdependent parts working together cooperatively to produce effects that would otherwise not be achieved. This construct, once validated, systematically repeats the performance under similar circumstances. In recognizing that the product of such relations is work, we go further and invent such combinations, and combinations of combinations to work to our benefit, calling each discrete compound mechanism a machine. We even use the term 'mechanism' descriptively to suggest that natural, even living, processes appear to function mechanically.

In making such a connection between the living and the non-living we are not only noting that the machine is suggestive of a living force, but also the converse, that a living organism is analogous to a machine. We even borrow the adjective organic to describe a mechanical arrangement of parts in order to pictorialize the manner in which it operates. In this way we freely borrow from one reference to describe another, and from the other to describe the one. Language allows us the latitude to call some thing something that it is not. We become careless in making literal distinctions between, for example, a machine and a biological organism. We say that a thing is organic if it displays characteristics that resemble those of a living plant or animal, as having a form that grows out of inherent factors that we loosely identify with inherited factors. The connection between the two, whether understood or subconscious, is often the fact of their both being compounds of carbon and hydrogen, whether derived from the living or not. An organism is thus a thing of many parts that functions, in part, in accordance with the dictates and interrelations of mutually dependent parts. By such a definition an organism is a broader spectral band in bio-logic than life itself. An organism may be a functional part of a larger organism, as a living cell is to a body, without the necessity for it being self-sustaining. We have, thus, described an axiom of natural law whereby events occur as a result of antecedent causes acting in sequence or combining to act concurrently in unison. An organization more broadly forms a set of relations, while the set of all organizations describes cosmological relations comprising the universe.

Systems

We use the term 'system' to identify a pattern of behaviour of interrelated parts forming an organic or organized functional unit. Natural laws infer preferential patterns of relations and behaviour and prescribe their limitations. The most alluring systems are those that work to perpetuate their own continuity. We have earlier tested the waters of habitat and evolution from which we distilled the necessary conditions for life to develop and persist. Time and changing relations have compounded to form seemingly infinite organic variety, each species being inhibited by the limitations of its habitat or by competitors for its material or spatial properties.

The study of the interrelationship between organisms and their environments that determines whether species of plants or animals thrive, survive or demise, we call ecology.

Ecology

We have travelled from simple ideas, order and organization, to infer the most complex and confounding of interrelations: organicism and life. We have belatedly constituted the discipline of ecology as that branch of science concerned with the interrelationship of organisms, the dependency of each upon its environment, and enquiry into limitations that govern the viability of species and populations due to natural fluctuations. We shall show in a later section how the economics of biomes or modes of life become unnaturally corrupted.

Since the logic of economy extends to all things as the overreaching system governing the operation of the universe, ecology may be recognized as being as much a branch of economics as of the life sciences. Conservation is a natural outflow of poverty, a process of economic moderation between supply and demand, from the benefits of which we build ethical constructs.

Ninety nine per cent of the air that we breath is excrement: 78 per cent the microbial waste we call nitrogen, and 21 per cent oxygen, the by-product of photosynthesis.

The soil upon which landlubbers depend for their ionized minerals and edible flora and fauna is composed, or more properly composted, in large part, of residual waste comprising organic and inorganic excretions and decomposing life forms. Finally, the large fishing holes, otherwise known as oceans, are the sink holes for all that flows and falls into them, adding to their copious stocks of indigenous organic detritus. From the above we come to the bitter-sweet conclusion that the fuel of life is life's excrement and what we presume to be nature's inefficient ways are not so, but rather super-efficient means by which all is waste and nothing wasted. It is the coordinated correlation of all with all else that is the secret underlying nature's ability to continue doing what she does unabated: evolving the architecture of all things and everything.

As to whether the earth is a self-sustaining organism or not must remain an open question; suffice to say that it comports in many ways to suggest that it is, if not alive, diabolically metabolic in its behaviour. Live or not, the earth exhibits some startling displays of self-control in the manner by which it regulates, for example, the climate and the chemical composition of its constituent parts. If not living, it is so simply because it does not conform to the way in which we elect to define life.

The reason that we have cause to care about special relations is that, in the final analysis, all relations are special. That principle cannot be better illustrated than by reference to the interrelations between living organisms.

Consistent with our emerging understanding of ecology, and in our own interests, it is imperative that we continue to seek an understanding of the principles that underlie nature's behaviour, as scientists and, more importantly, as the culture of the dominant species, the one that knows all others, mankind. In this way, and only in this way, can we hope to avert generating disasters that grow in proportion to our expanding body of knowledge and consequent capacities. To this end it is the generalist rather than the specialist that we must look to for generic solutions to our present dilemmas.

The interminable flux of all things physical and chemical beget symbiotic relations that somewhere, sometime, sooner or later inevitably give rise to things self-sustaining and biological. By extension and the same application of logic, probability eventually delivers the capacity for biological systems to address the biosphere logically, and to introspect upon the significance of doing so.

53 Life: Progenitive Vitality

What constitutes the threshold between the exclusively physical and chemical on the one hand and the biological on the other is human definition. But how do we make our distinctions between the living and the non-living? We look for commonalities in behaviour. We look at what all living things do as a class that the non-living do not do. Class membership is our critical threshold.

As with so many enquiries, it is helpful to explore the five cardinal questions 'what', 'where', 'why', 'when' and 'how' in investigating life. By taking this well-trodden path we can start, proceed and anticipate knowledge, however tentatively.

What

So 'what' do living things do for a living? What do we mean when we use the word life? We have witnessed the preference of gases, liquids and solids to congregate with their own kind in the atmosphere, oceans and soil, admixed as they may be with lesser concentrations of each other. But there are peculiar and related combinations of matter that form their own distinct integrities, the essential characteristics of which appear to be few in number, perhaps just six. They comprise 'formation', the coming to being of a living organism; 'sustenance', nourishment through ingestion; 'development', formal change through growth; 'vitality', systemic potential for acting in self interest; 'procreation', the reproduction of kind; and 'mortality', cessation as a viable system. A timely subscription to all of these possibilities constitutes membership in the realm of the living. That's life!

Where

Science has not reached its conclusions as to 'where' life started so we are free to guess. Did life start somewhere along the line of confluence between the earth, ocean and atmosphere, or deep in the warm belly of the earth to be leached out to the surface or breach-born through volcanism? Was it sucked in by gravity while cruising in the cosmos, or were its origins in the deep troughs of the oceans where hydrothermal vents accelerated the flux of chemical activity and provided a wide range of thermal regimes potentially optimal for the formation and maintenance of life?

Why

'Why'? Because the right conditions for beginning and for changing prevailed. In the potpourri of swirling elements that comprise the earth, sooner or later any element will come into contact with any other or any combination of others, simply by sweating out the rule of probability. By this means we arrive at a critical mass just 'yearning' to be something other than what it is.

When

Estimated dates for 'when' the origins of life on earth first gestated linger around 4 milliard (4 US billion) years ago, a time when the earth was likely much hotter than now, the atmosphere heavily laden with carbon dioxide but very light on oxygen, and when the oceans swirled elementary and compounded solutions of mineral waters. The answer to the question 'when', when applied to change, is whenever conditions change critically. Since they are in a constant state of change, a string of critical transformations occurs. Hence evolution.

How

There is always a big debate on any question concerning 'how'. The metamorphic path to life remains conjectural so let us conject for a moment. It would appear reasonable that, in the natural tendency of relations to become more complex whereby dissimilar inorganic matter forges unions that react chemically, the outcomes of such unions would be more or less pronounced dependent upon prevailing conditions. There would be a higher probability that change would be facilitated, and with greater expedition, if additional energy or matter were present that would act as a mediator, catalyst or accelerator to a reaction.

A smidgen of clay is kissed by a drop of water, which it absorbs. Pregnated it expands. The sun rises and drives the two apart and we are back to ground zero. When the sun sets the water condenses and they get cosy again, just like in real life. Do this for a few million years and compress the idea in time for better comprehension and we have the 'birth' of a reciprocating mechanism. Like kinds of matter migrate towards each other, as we have seen, to bring a continuing supply of materials for incorporation, can we say by ingestion? Add a pinch of colourful heat-absorbent pigment, say chlorophyll, and the action accelerates, the engine revs up.

The more chlorophyll the greater the speed. Allow a few newcomers to join in, ionized minerals perhaps, and we have growth. Shake the cocktail of building blocks and we have a new order, a development, no less. Insofar as it continues along this course it is doing what it 'likes' to do, looking after its self-interests, all in conformity with the principles of inertia. Isn't that an act of vitality? It breaks in half under stress and the two parts, lightened by the release from their ties, are free to continue doing what comes naturally until their critical masses slow them to breaking point again. *Voila*! We have a simple form of asexual reproduction, and baby alternators, or are they single-molecule micro-organisms of DNA called procaryotes or bacteriochlorophylls on their way to becoming single-celled organisms? The party may swing for a long, long time but sooner or later a cataclysmic event will rend them asunder, as when the actions of plate tectonics draw them through subduction into the fiery world of volcanism, consummating the death of a great idea: life.

The six criteria for entry into the class of the living appear to have been satisfied. If this course to how life may have started doesn't meet with your acceptance, think of another one, your progenital path. The extraterrestrial microbes locked into an earth-bound meteorite theory may be plausible but begs the question as to how 'they' acquired life status.

Getting a Life

Let us take a walk for a moment on the wild side of life. Imagine that a flake of chlorophyll participates in the first live experiment. Remember that it absorbs heat by which we can understand that it expands with the possible effect of shielding the most sensitive of evolutionary events from the harmful effects of infrared light. When not exposed to infrared light it contracts, allowing the surface of the little 'it' to be exposed to other influences. Functioning like a thermostat, the chlorophyll senses and reacts to the presence of infrared radiation, its actions having the same effect as a rudimentary neurological system in protecting the emerging organism. A chain of evolutionary life-promoting events may have followed from chlorophyll to chlorophyll-a that absorbs visible light, to cyanobacteria that engineered the splitting of water to access hydrogen, inadvertently precipitating the chemistry of photosynthesis. The release and accumulation of free oxygen, the emergence of crisis conditions for photosynthesizers due to critical changes in the apportionment of nitrogen and oxygen in the atmosphere forced some forms of life to adapt or become extinct. To be or not to be oxygenated, to use it or lose it, or to become denitrifiers, those were their questions. In so doing the processes of life's ancestral precursors were reversed, by which means an approximate status quo was achieved between the production of oxygen and replenishment of nitrogen. The burgeoning production of oxygen enabled the evolution of multi-cellular organisms. As the levels of oxygen rose, nature's counter-force of oxygen-gobbling respirers evolved. By now the family tree of life had many branches, one of which would eventually lead to you.

Green Things

The most fundamental question concerning the continuity of life is how raw energy is captured, tamed and converted to become a domesticated life force. We shall start by examining nature's solar energy collectors, since this single mechanism is crucial to the development of almost all life forms as we know them today.

Photosynthesis has two primary aspects: the collection of high-energy photons, which prompted the 'photo' prefix to the name, and the harnessing of that energy to convert molecules of inorganic matter into organic building blocks, accounting for the 'synthesis' suffix.

Captured by the heat-absorbent pigment chlorophyll, solar energy is applied to partition water into its two constituents, hydrogen and oxygen, and is compounded into a storable chemical form for 'domestic' use. Typically the photosynthesizers draw carbon dioxide. The hydrogen of water is fused to the carbon of carbon dioxide to form organic carbohydrates comprising cellulose and other vital compounds used in the maintenance and building of new cells, while the oxygen is discharged as unusable waste. The reserves of 'domestic' energy are drawn upon to drive the reactors that break down carbon dioxide, to synthesize sugars that power the transportation of nutrients to the building sites, or for physical movement.

The reactors alluded to above are enzymes that promote linkages, in this case regulating the relative mix of the carbon of carbon dioxide and the oxygen of water through the auspices of a mediator enzyme, rubisco, that facilitates the intake of carbon dioxide into the molecules of plants and algae. With a regulatory process in hand, the quantity of carbon harnessed can be controlled to relate to the availability of oxygen, or vice versa.

What determines the rates of photosynthesis? In large part the surface areas of the solar energy collectors, subject to the total flux of sunlight available at a particular site. The photosynthesizers have a canny knack of maximizing the surface areas of their boundary membranes exposed to light. Ecologists measure these areas for each biome as ratios of the surface areas of the earth that photosynthesizers occupy. By these gauges, called leaf area indices on land, the index for the evergreen broadleaf rain forests of the tropics is a factor of about nine times the surface area of the land. The ratio graduates down to zero in desert and polar regions and averages about 3.5 times the land area of the earth or the equivalence in surface area to one planet earth.

Similar measurements, when applied to the surface areas of marine photosynthesizers, yield aggregated results in the order of six global surfaces. Marine photosynthesizers are microscopic in size, their number greater and their surface areas larger in proportion to their mass, all of which work to offset their relative inefficiencies due to overlaid water.

The photosynthesizers on land include broad- and needle-leafed plants, grasses, mosses and algae, while the ocean harbours diatoms, coccolithophorids, algae, cyanobacteria and seaweed, including the giant stands of kelp along shorelines.

Through the aegises of chlorophyll and rubisco, photosynthesis may be said to be the principle action that facilitates cell construction that inevitably leads to consequent atrophy, providing the realms of atmosphere, ocean and soil with their common matrices of biotic matter.

Unlike other forms of life, photosynthesizers do not extract energy from their food; to the contrary, their needs are for low-energy inorganic nutrients. Photosynthesizers may be said to begin the food chain by virtue of their converting the inorganic to organic matter through harnessing the energy of the sun. Notwithstanding impressive leaf area indices which exhibit great surface areas of foliage, only about 0.1 per cent of the solar energy reaching earth finds its way through photosynthesis into life, the balance being reflected and absorbed by other branches of the possibility tree. Therefore, with respect to energy cycles, life's diversionary influence is minuscule.

Groovy Movies

Vegetarians by circumstance or choice are second in line consuming the products of photosynthesis directly to gain matter and energy, while the carnivorous rely for their material and energy gains upon eating the flesh of others who have already synthesized just what the carnivore needs. From the foregoing it should not come as a great surprise that the identity and apportionment of chemical elements in consumers closely resembles that of the consumed. As a fact of matter, 'we are what we eat', as Winston Churchill succinctly put it.

While fully dependent upon the good services of the photosynthesizers, the respirers of this world function to reverse the efforts of the photosynthesizers by inhaling oxygen and exhaling carbon dioxide. In so doing they contribute the counter-force that compensates the atmospheric equation.

A symbiotic relationship exists between animal life that inhales oxygen and exhales carbon dioxide, and plants that effect the reverse. In this way, a self-corrective balance is maintained between the two, each placing limits upon the other as the primary source of the essentials that the other needs. In the case of lichen, they comprise a coalition between fungal respirers and algal photosynthesizers whereby fungi acid-etch the surface of rocks, while algae feed upon the mineral nutrients so exposed – to their mutual and collective advantages.

The duet involving photosynthesis and respiration displays yet another manifestation of the essence of nature's not-so-secret strategy: to counter force with force in gyration to the effect of approximating neutrality.

The final step in the loop rests with the decomposers, fungi and hyphae in soil and bacteria in the oceans, that work to salvage and restock nature's base materials from biotic waste, excreting enzymes that break down organic matter into inorganic ions, thus making them accessible to photosynthesizers once again. Bacteria, fungi and root cells release carbon dioxide and consume oxygen. The carbon dioxide percolates through the soil and is released into the air. It is noteworthy to reflect that the decomposers exhibit the same propensity for maximizing the surface areas of their boundary membranes as photosynthesizers in order to enhance the efficiency of their counter-culture.

Reciprocity

The concept of waste is relative. Each organism takes what it needs from its environment and discards what it cannot draw benefit from. That discard, whether passed by or passed through, is in demand by others, larger or smaller, with differing needs and capacities for utilization. This peculiar reality is not confined to the organic world; indeed it might be said to be characteristic of all matter insofar as all matter has native proclivities for establishing affinities, whether they be chemical, physical or biological. Affinities, one might almost say affections, and their antipathies, seem to be the most fundamental and powerful drivers of change, forming the boundary conditions by which we make distinctions between the living and the non-living.

Could it be that life is somehow reciprocating the favours afforded it by the non-living, each serving to hold the balance of nature approximately constant for the benefit of all? The test of such a hypothesis would be to compare earth with life abounding to earth without life. An earth without life would degenerate to a point where methane would not exist, and oxygen would diminish to trace quantities.

What, we might ask, constitute the common threads of life's existence? Lines, Aristotle once suggested, are a successions of points. To the extent that metabolic threads are the paths of individual atoms and molecules metamorphosing their way through the living, perhaps he was right. Metabola (from the Greek word *metabolos*, meaning changeable) are linear and directional.

Metabolism is a highway system for convertibles, from solar energy through water-splitting photosynthesis to chemical energy, via carbon dioxide and water to carbohydrates, where oxygen exits as waste. Anabolism is the energy-guzzling vehicle that transports molecules from minor to major status (e.g. amino acids to form proteins) in the manufacture of building systems and the construction and maintenance of vast communities of cells. That they do this automatically against all odds (well, countless billion-to-one odds) speaks of some kind of predisposition (can we say 'will'?) to reproduce prolifically. Are amino acids alive? Is this commitment motivated by what Plato called the strongest of all desires: immortality? Not quite, but they appear to function as indispensable orchestrators of life.

Across the median strip the counter-flow of catabolistic traffic is bound to convert large molecules into small (e.g. glycogen to pyruvic acid) by which means chemical energy is conserved to drive other metabolic internal combustion engines or be converted to physical energy. Metabolism can be said to be the combined actions of boundary crossings, without which life cannot sustain itself. Such actions lead one to draw conclusions regarding processes that cannot be reached by studying each boundary exchange in isolation.

Jellies (jellyfish) are the clearest visible expression of what life is all about; they eat, transfer, excrete, breath, metabolize, move, reproduce and respond to their environment in the most direct manner, with no apparent redundancies. They are the apogee of what the organic architecture of other realms should be in their simplest expressions, to which we aspire with great difficulties.

Number Crunching

On average, the human body is estimated to comprise approximately 100,000,000,000,000 (10^{13}) cells of which about 98 per cent are replaced in the course of a year. If we divide 98,000,000,000,000 by the 31,557,600 seconds in one year, we will recognize that, on average, we are adding and subtracting over 3 million cells to our bodies every second of our adult lives, each carrying a replicate library of the host's inventory of genetic messages. At about the age of 30 the ratio of cells gained to cells lost shifts from a surplus to a deficit.

Each acquired replacement atom is recharging us with some kind of qualified quantity (energized matter). For example, iron's passage through life may be observed in the pigment haemoglobin that, bridging distinctions between vertebrate (animal), microbial and leguminous plant life forms, affords a common molecular path to exotically variable ends. Iron assists the common needs for shunting electronic energy where it is needed for photosynthesis and respiration through the mechanism of ATP (adenosine triphosphate), the energy carrier in living cells. Similarly, calcium serves as a structural element in building plant cell walls and bone formation in vertebrates.

We can infer that a significantly greater or lesser number of atoms would result in increased or reduced levels of functional performance. Through experience we can approximate ratios of intake to functional performance (output) and establish the thresholds of dysfunction at the high and low levels of quantitative intake. Through this understanding we have arrived at Descartes' notion of the human body as a complex mechanism. What we are addressing, however, is the constancy of a systematic whole undergoing change. You are not simply the sum of your substantial material parts at any point in time; you are, in addition, a complex set of insubstantial relations, each of which is interacting with the others.

Making Connections

If it is relationship that links matter to matter, then it can be inferred that matter links relationship to relationship. Insofar as we claim to understand matter, we do not have the same degree of security in our knowledge of relationships, the tangible being more accessible to investigation than the insubstantial by virtue of our primary reliance upon sight, our dominant sensory system. One does not 'see' a relationship between bodies. However, intangible forces have their own logic that we are simply unable to *tange.*

Further, if we recognize that each atom of each cell is a distinct entity that is discretely separated from each other atom, and that each electron is separated from its nucleus, we must conclude that each of us is some kind of temporary and loose arrangement of migrating atoms, a force-field or animated cloud. Who are you and who are you going to be in half an hour? We are space-time-events that, because of the particular way in which we classify experience, we think matter!

In life, confirmed by human experience, what an organism can or cannot do are equally governed by its attributes and deficiencies. The more fundamental the attribute, as with say a gene, the more finite the presence or absence of a particular faculty. As in mathematics, development occurs only from a point of existence, from which performance may be enhanced.

From a point of non-existence one can multiply the value zero by any factor and no development will occur. In life it takes an anomaly by mutation to develop six toes on a foot, as occurs not infrequently with the Waorani people of the eastern slopes of the Andes in Ecuador, from which point genetic inheritance may proceed.

Individual genes are compounded elements and molecules that function like switches to pass or deny the passage of energy. When 'open' they cause connections to be made, the effects of which are to facilitate interaction. Thus, a particular combination of genes will trigger a particular event or tendency, which, when exposed to its environment, will result in a particular set of mutual impacts.

Doing It

We return to the question what does life 'do'? What does life contribute to the greater scheme of things and vice–versa? All life forms draw from the atmosphere, directly or indirectly. In doing so the living act as pervasive pumping systems, stirring the atmosphere much as the gyrations of the earth and convection due to temperature do, but more so. More so because life adds, subtracts, multiplies and divides the constituents of the atmosphere, mathematically speaking, and in so doing tempers the earth's excesses. Doing is all-important, being causal to subsequent events. Life promotes chemical changes that would otherwise not occur, as in the liberation of free oxygen that enables the respirers of life to prosper and further develop. And life accelerates the oxidization of minerals along their courses towards their simplest ionic forms, thus making them accessible as nutrients to recycle and serve life again. Life generates the mantle of carbon dioxide that shields sensitive forms from the ravages of the sun's ultraviolet light, which would otherwise make life as we know it inviable.

Upping the Anti

So life is nature's accelerator, exerting energy in opposing directions, speeding up the rates of change, always moving from the simpler towards the more complex organic forms. And what do more complex organic forms do that simpler forms cannot? They evolve, or invent through the genius of their leading exponent, man, new ways of further accelerating the pace of change.

What purposes are served by this compounding acceleration? Why is this so? The 'purposes' (suggesting but not necessitating will) are those interests that are served by and benefit from opportunities that flow from such changes, and also those events that are no longer rendered possible. The effects 'are' the purposes, nature's purposes, a point from which we may slip uneasily from fate into fatalism.

Keeping It Up

From the viewpoint of life, it is life's urgent purpose to continue living, and its emergent purpose to develop, that are served by change. The 'why' is obscure, perhaps even non-existent unless one wishes to attribute inertia or invoke metaphysics as the wilful cause of an as yet undisclosed plan, or to attribute the cause to anything else on an individual basis for one's own peace of mind. But one thing appears to be so: all change is facilitated by energy; directly from the chemical reactions of stars of which the sun is our patron saint; radiating to overcome separation, conducting where in adjacency, or convecting where free to flow – all provoked into frenzy by the existence of matter. While the 'wills' of all individual parts of the universal system appear to opt for the tranquility of inertia, their relatedness queers their natural dispositions towards equilibrium and moves them reluctantly into the restless mode of doing. Thus, as long as we have plurality of entities we will continue to have change.

To more fully understand human life it is of primary importance to distil its essential nature. That essence extends beyond considerations of relationships between identifiable bodily matter, and beyond the kinds of bodies that act to link types of relationships. What human nature incorporates, and what our collective understanding needs to grasp, is the nature of incorporeal, insubstantial entities such as consciousness and ideas. From such understandings we can then address 'their' relations to matter and to relations between matter.

Changes in form, function and relations imply passing on, but to what effects? Human life is the process by which we graduate from a state of believing that everything is possible to a point where nothing is possible – when we die.

54   Death: The End Game of Life

Death by Natural Causes

Ageing presents itself almost imperceptibly. It manifests itself without pain as a very gradual loss of capacity to function at levels of performance formerly possible. It is attended by a dulling of the senses, of which the loss of joy is the most predominant. Distinctions in time, place and daily events dissolve into monotony. Melancholia displaces interest as the passions are slowly extinguished and the intellect dulls. The body atrophies. Earlier promises of a welcome relief from the burgeoning demands of living and the prospects of reflection and the joyful giving of counsel are transposed into a uniformly colourless continuum of empty, almost unbearable tedium. The greatest loss is of that former inner fire, that knowledge of personal potential, that sense of mission that slowly becomes extinguished.

Something very sad happens when people get old but before they become senile. They become ultra-conservative and paralyzed by fear of change. It is as if they have forgotten all the stimuli in life that prompted curiosity, the spirit of adventure and that sparkle of love for life that is its essence. Death by natural causes starts in the mind. What triggers it is complacency, that relative satisfaction with what one has compared with what one might not have if the winds of change were to blow in unfavourable directions. At such times all winds of change and all directions appear to be unfavourable!

Notwithstanding the natural course of bodily entropy, the environment aids and abets the process. Disease attacks the fabric of the living, rendering it less than able to resist the ravages of impairment. Peace of mind becomes the ultimate and eventually the only desire.

The Degeneration of the Mind

All changes in time. In ageing, the impetus to advance into new, green and pleasant pastures wanes. Any progress into unfamiliar territory seems to require a measure of energy disproportionate to that that formerly enabled advancement. The efficiency of the life engine diminishes. A slowdown of mental functions succeeds.

As the brain loses its integrity, vital contacts disengage and random short-circuitry ensues. So the perception of a world to which the mind has become accustomed metamorphoses into otherworldliness. With increasing frequency nobody is 'at home', or if they are, the visiting hours and the passwords are not posted. From outside the 'box' stupefaction appears to be taking possession, but inside the movie continues. Unlike the ordered sequence of ideas necessary for comprehension, the memory is denied access. A change in process from a selective to a random generator gradually takes command of the organization and direction of all mental assets. We can no longer make a logical distinction between what is happening and what we want to happen. We enter a world of random imag(es)-in-a(c)tion. In such an environment, anything is possible. Through the agency of hallucination, gods and other fantasies are conjured up to colour death, as a positive, even beautiful prospect.

Spirituality is the default mode characteristic of minds under stress whereby defective performance due to loss of integrity may be brought into conformity with the law of entropy that maintains continuity in the distribution of influential parts. Insofar as we do not live in the past or the future, and recognizing the present to be ephemeral, life and death become indistinguishable, as dreams are to nightmares, while living the succession of moments is still very real.

Death by Misadventure

Life is a serial play of risky events. Those who are naturally disposed to seek high goals and to extend their self-knowledge into realms of the unknown are particularly vulnerable. The pursuit of knowledge is inherently dangerous because it exposes one to choices the consequences of which are beyond our ability to fully comprehend in advance. It is the lot of optimists to voluntarily carry this burden of risk in return for which lie all the prospects for improvement of the human condition. Without the advent of risk the body and mind stagnate. Stagnation instigates disintegration. In either case, death comes by default. But how much sweeter are the fruit of having tried and died than of having denied and been denied!

Death by Design

Nature predicates killing upon necessity. The animal kingdom is ruled by the law of the jungle. One species of animal, mankind, has advanced to a state of self-awareness whereby he raises questions concerning such necessity. He finds himself part animal and part god, unable to clearly distinguish when he should act as one or the other. Therein lies his most conspicuous dilemma.

Homicide, the killing of one person by another, is murder. Execution as a legal penalty is state-sanctioned murder. War, the state-sponsored killing of many by many others, is murder. In each of these cases man relapses into his animal mode of behaviour, relying upon his savage emotions to authenticate his actions, but with a distinct difference – he does so with premeditation, with intent to kill without necessity.

The ethical question presents itself large as to whether the advocacy or commissioning of premeditated killing by elected or appointed officials, or the failure to commute capital to life sentences, should be automatic grounds for punishment rather than exemption from otherwise normal social consequences, thereby placing them above the law.

Suicide, the taking of one's own life, is a matter of conscience, the private conscience where an individual does not want to burden others when he can no longer care for himself, and the public conscience where social conditions drive an individual beyond hope to despair. These are natural events that overreach any legislative remedies that society may contrive to regulate. There is an element of hypocrisy attached to outlawing suicide on the one hand and in permitting the degeneration of social conditions that cause such despair on the other.

Euthanasia, the practice of putting persons to death painlessly to relieve suffering, is exposed to moral objections on the grounds that it undermines respect for human dignity, contravenes the practice of medical relief from suffering, fails to resolve ethical questions concerning whose judgement is involved, invites dangers of abuse and denies opportunities for fulfilment for those that are so afflicted and those dedicated to serving the infirm.

Almost every abortion, and denial of abortion, whether legal or otherwise, is the direct or indirect consequence of man's callous disregard for the interests of women.

All of the above practices corroborate a pattern of inconsistency and instability in human values, decisions and actions that cannot be reconciled with the all-embracing idea of civilization.

Death as a Cultural Obsession

The most conspicuous proponent of the 'good life', the United States of America, carries the greatest responsibility for the promotion of a better world by virtue of its dominance as an economic, political and military superpower. While the high-minded values engendered in the US Constitution may be succinct, the means by which the US often proceeds to secure its espoused goals are profoundly disturbing. In the United States the promotion of death is big business. Why? To ask this question is to answer it. The question and answer are inseparable.

The US military establishment is the largest subdivision of the federal government and commands the largest discretionary budget. Americans rally around their president in times of war. The application of military force thus offers favourable prospects that political ambitions may be rewarded, all other considerations being equal. This is reflected in the fact that every president since World War II has embarked upon overt and/or covert military operations that, beyond their obvious designs, have been calculated to enhance their political reputations, even though, more often than not, such military and political campaigns have resulted in failure.

By this doctrine we identify presidents Truman with Palestine and Korea, Eisenhower with Lebanon, Kennedy with Cuba, Johnson with Vietnam and the Dominican Republic, Nixon with Vietnam and Chile, Ford with Vietnam, Cambodia and Angola, Carter with the Iranian hostage crisis, Reagan with Grenada, Nicaragua, Afghanistan and Lebanon, Bush (Sr) with Panama, Iraq and Somalia, Clinton with Haiti, and Bush (Jr) with Afghanistan, Iraq and (through proxies) Lebanon – a less-than-complete catalogue of inauspicious commitments.

One cannot but reflect upon President Reagan's representation of the Soviet Union as an 'Evil Empire' and wonder whether the US has not become the mirror image of that bombast. Abstracting from the lyrics of *The Dark Side of the Moon* by the rock band Pink Floyd, 'The lunatics are on the grass' seems to fit the case, as if those who so make-believe and act upon such beliefs must be under a narcotic-like influence.

This affliction, an outgrowth of the Western fantasy that 'might is right', grips the imagination of Americans to the degree that the primary thrust of the news and entertainment industries is directed towards, if not by, violence as a justifiable means to securing coveted ends.

Clearly the US president should not have any authority to declare war; an act so grave should be the sole prerogative of Congress, and then only by a 66 per cent majority vote in favour.

Many cultures foster conflict in the name of progress, for which mankind suffers dearly. Wilfully killing people is, by any standard, a very strange, deranged form of service or servitude. The examples cited are perhaps made more incongruous given the virtuous intentions upon which the US nation was founded and, in all fairness, frequently rises to reaffirm and demonstrate.

Are we not entitled to believe that it is not too late to engage in the creation of the opposite, a sweeping utopia where none will be able to decide for others how they die, where love will prove true and a measure of happiness shall be accessible to all? You don't have to be insane to kill people, we have large standing armies awaiting instructions to do so, but it helps!

The Civilized View of Death

Imagine circumstances wherein there would be no worries, an end to pain, no obligations and no pretence. Such a condition would be free of responsibilities, hardship and harshness, free of the burdens of work or the necessities for routine. The whole prospect of negative experience would be stripped away, leaving, just leaving. All those relations that one may wish were not so, would not be so. Such is the outflow of death.

The great religions have words for such states of relief. They are presumptively given place names, as if any place isn't good enough. But all places are any places and all places are such places. For it is not for want of knowing that one is here rather than there. One is here if one is there, or anywhere, because 'here' is by definition where each of us is at all times, and 'there' is where other people are. So if such circumstances as we have outlined exist, they must be here. The season of death may be untimely, but be assured; the remission of hardship will present itself. As with other states of sublimation, anticipation may be almost as poignant as the actual event. So we have every reason to be happy and to welcome a beautiful prospect. The Promised Land awaits all.

Compassionate man (about whom we shall have more to say), in his relations with those less fortunate, seeks to anticipate and facilitate relief for those in distress. Through his example he engenders support and cooperation, by which means others do as he and so set a pattern of exemplary behaviour in counterpoise to the culture of killing.

Obituary Thoughts

If, instead of waiting to a date too late at which time someone may or may not deign to write one's obituary, one were to write one's own obituary early in life, amending it as circumstances dictated, how much more pertinent it would be to that life. How well it might serve as an incentive to dictate the conduct of that life in the knowledge that, if fulfilled in the manner prescribed, it may serve as a passport to immortality.

Epitaph

> Somewhere
> Where my heart may rest in peace
> And my mind from torment cease
> Where blue and green
> Are king and queen
> And the gotten are forgotten.

55   Mankind: Society in the Aggregate

The distinction between man and his society is that between the singular and the plural. When we are alone we readily recognize that we are responsible for our own nourishment, shelter and security. In company we generally respond compassionately to urgent and critical needs of others for such essentials. It is when we are neither alone nor confronted by emergent and threatening circumstances that we are less certain about making and sustaining commitments. Under such conditions necessity releases us from the bind of its spell and we are left in a quandary as to who is, or should be the master of our conscience and the governor of our actions. This is the twilight zone and the purview of politics, the essence of which is the definition of shared interests and the promotion of social goals. The first rule of society is to protect its citizens. The second rule is to secure opportunities for responding to other essential needs. The third rule is to accommodate change within continuity. It is no coincidence that these are also the primary rules underlying individual behaviour. Each is behoven to the other to reconcile differences.

Alone we have no personal relations. When we associate with others we have. First, they are implicit, the bases upon which we tacitly agree to continue associations, and later verbal or written statements that sanction and define the nature of such associations. Sanctions are in the nature of accords that formally endorse relations between people and licence entitlements. If one is born into a society and benefits from its collective advantages one assumes citizenship, which citizenship carries reciprocal responsibilities. If the perceived 'benefits' exceed the 'costs' we voluntarily retain membership, as of a family.

Sovereignty

Alone, the first person singular is king. He ordains what shall be done, and where, why, when and how it shall be done. In the natural family, tradition has held that the man, being physically stronger, acts in that capacity. In the tribe the elders serve to advise the chief who reigns supreme. In the monarchy the court advises but the king is king. In totalitarian societies the dictator dictates. In each of these patterns of social organization absolute authority resides at the pinnacle of power. The prospects for wise or foolish governance of the many hinge on the judgements of a single soul. To remain in power that single soul must be all-powerful. He can be so by the will of the people or by brute force. To remove him requires a greater will or force. Civil disobedience, revolution and war thus become essential contingencies reserved by the people for the people. The people hold the keys to ultimate authority, but the keys are locked away in contingencies that they are very loath to invoke.

A good government, according to Plato, requires a balance between two principles, *eleutheria* (popular control) and *monarchia* (personal authority). Communism is rejected as impracticable in a society of ordinary human beings.

As we have shown, ultimate authority rests with the people. If this is so, why should it ever be necessary to resort to brute force to retain what is naturally theirs? What does it take to remove tyranny of a majority by a minority? It takes government by a majority. This, it may be argued, may simply invite tyranny of a minority by a majority, but it cannot be successfully argued that government by majority is not incrementally more just. This predicate is the foundation of the modern republican form of government.

Alone, the first person singular is a majority of one. In the natural family experience has shown that each of the parents has unique contributions to make to the family unit, dividing authority on the basis of natural capacity and agreement. The modern tribe institutionalizes its authority through incorporation so that its form of government may bridge the disjunctures caused by mortality. The modern monarchial form of government closely resembles the republic, while retaining its monarch and privileged aristocracy as token expressions of continuity within traditions.

Government by majority has some practical limitations, not the least of which is the logistical problem of getting all eligible citizens to meet and vote regularly. As the state grows in population and geographic area, each citizen becomes commensurably detached from the centre of government until it becomes necessary to govern by proxy, delegating authority to representatives to vote on issues on behalf of constituents that share their general values and goals. The question then arises, how big is too big? The path to a satisfactory resolution of this question lies in the distribution of power proportionate to the interests of those affected by each kind of decision. Thus, personal issues are considered and acted upon on an individual basis, local issues are addressed locally, regional concerns attended on a regional basis, and only those matters that bear upon the interests of the whole population, including relations between the state and subdivisions of it, should logically be the legitimate business of the central government.

Population

Approaching the close of the twentieth century the rate of population increase worldwide was about 1.8 per cent per year (down from 2.0 per cent in 1970). For the then population, brinking 6 billion, this translated to an annual increase of 108 million people, equivalent to a new city the size of New York every 25 days, or a new state the size of Mexico each year.

By this account an event comparable to the bombing of Hiroshima, where 60,000 lives were lost, would offset the rate of population increase for a mere five hours, and to World War II, when 60,000,000 lives were lost, for just 200 days.

Without going into the logistics of support, it can be seen that this phenomenon rapidly reaches crisis proportions. If this rate of increase was to be maintained the numbers would compound like interest in a savings account until, a little before the year 2600, there would be one person per square metre (or square yard) of land surface on earth. Upon learning of this projection, architect-philosopher Paolo Soleri, a tireless advocate for redressing the problems inherent in accommodating burgeoning populations, responded with a twinkle in his eye: 'Standing room only!' Upon reflection, a span of six hundred years only takes us back to Geoffrey Chaucer!

The twentieth century's fanatic obsession with economic growth, predicated largely upon projected population growth, is only valid as long as growth is maintained. It can be readily recognized that in many parts of the world the population has already exceeded the ability of the environment to sustain it. The real dangers to democracy are two-fold: where economies are underproductive, populations are driven to sustenance levels of existence whereby their only interests are short term, and alternatively where economic growth is excessive it places large populations in a position to make excessive demands for frivolous material possessions. Both place unsupportable demands upon natural resources. A balance between these extremes, involving a more equitable distribution of wealth, requires a humane means by which the rates of population growth can be lowered and eventually stabilized. Man, as the dominant species on earth, bears the burden for safeguarding the viability of life on earth, his means: personal conduct and responsible representative government.

Ideal Forms

Insofar as one's individual values and goals correspond with those of society, and one has a voice in the formation of laws that uphold these values and pursue these goals, we can say that the relation between man and society is ideal. But in a pluralistic society values and goals differ, as a result of which compromises must be forged between divergent interests. To the extent that compromises are founded upon common denominations of value, it is the more popular views that prevail, one might say the median views of ordinary people. Mediocrity is the standard by which all social issues are judged. The views with the greatest potential for upsetting the status quo are cast out by the process described above. What, one might then ask, is the price to be paid by society for the loss of value of the extraordinary ideas of extraordinary people? The burden of persuasion falls upon the advocates of unconventional views who, only through dissemination and sweetening their proposals with modifications, can hope to secure popular acceptance. Those propositions that are clearly detrimental will fail due to their own inertia and no amount of sweetening can be added to attract the requisite support. By the same standards, a proposition of great potential merit requires critical measures of alacrity and persistence to persuade a majority to accept it. The burden again falls to the proponents to advance their cause through couching their arguments in terms that offer the greatest benefits to the greatest number of people. If the proposition fails, it will be for one of two reasons: that it does not merit acceptance, or that the arguments put forward in support of its adoption were not persuasive enough. In this latter event a better case may be presented at a better time. The doors of hope remain ajar for prospective improvements.

Concepts are invariably simplifications of form. They are so because they are discretely isolated from their context for the purpose of illuminating their central ideas. There is no better forum for illustrating this point than to retrospect on the simplistic notions that have gripped the imaginations of society in the belief that they had found their own particular ideal form. The early eighteenth-century concept of self-interest gave way to enlightened self-interest, then to the greatest happiness of the greatest number, and on to utilitarianism, the labour movement, the welfare state and classlessness; all sound ideas advancing justice beyond the preceding concepts, but none sufficiently all-embracing to stand the tests of stability and self-correctability over time.

Humanism, released from the shackles of its doctrinaire renaissance, naturalistic, scientific and religious prefixes, is nothing more or less than devotion to the spirit and practice of humanitarian kinship, and the reciprocity of support that it implies, to which the path of fuller knowledge leads.

Implicit in what we have discussed here is the idea that, in the ideal form of society, people 'should' be the determinants of their own forms of social organization. Inferred in this is the idea that the wills of the greater numbers 'should' prevail over those fewer in number. While this idea may suggest to the reader a democratic order, it does not preclude captive minorities, harnessed by majority assent, from being bound to do the bidding of the majority, to wit effectual servitude.

Social Improvement

It is the idealization of form that gives us insight into possibilities and a basis for defining social improvement. Consider, for example, the effect of scale upon the identification of people with their governments. If we hold to the view that direct participation in government by the people 'should' result in a more just society, then we should question the size of political territories as they bear upon that social objective, and tailor the territorial boundaries to optimize their functional utility. If, on the other hand, we conclude that the most just society is a society of one person, and that justice recedes proportionate to an increase in numbers, then we apparently cannot use justice as the sole determinant of the optimal size of political units. The same tests may be made with regard to optimal defensibility, economic wellbeing, administrative burden and other social necessities and goals. It is only when these criteria are compared, and only then as 'the people' so decide, that the prospect of an ideal model state can emerge, with the caveat that, as technology and society evolve, the gauges by which we assess optimal performance will of necessity also change, implying changes in conclusions to be drawn as to what constitutes optimal size. From a purely practical standpoint it is not desirable to keep changing political boundaries. People get very upset. So one is led back to conclude that the most secure entity, the most economically efficient unit, the most administratively complete element, the most just system, is a society of one: one world, politically subdivided by region and locality as we have above discussed.

State Patronage

Implicit in this subheading is the necessity to draw lines as to what should be in the public domain, what should be sanctioned by society and what should be left to private initiative. In making such determinations we remind ourselves that we are facilitating the future of others in a world that will likely be more complex and changeable than our own.

States and nation states come into being for the primary purpose of securing their territorial boundaries against unwelcome interference. Such interference comes from outside. The notion of a worldwide security system becomes quite appealing when we acknowledge that, under such a system, there would be no 'outside'. We, thus, have a military establishment solely concerned with defence, a kind of political doctor administering to the infirm to protect against the broader infection of the body politic. By the same measure we can readily appreciate that disease does not recognize political boundaries and that only by affording healthcare services to all, can all be protected. Guaranties of nominal standards for nutrition, shelter and education yield comparable social benefits.

Having addressed our personal needs we must then extend care to our habitats, the stock of plants and animals that nourish us, and whatever in the line of nature nourishes them. If we try to trace this line to its logical conclusion we find that it has no end, it loops back to include ourselves, and off we go again. It is therefore in our best interests to protect all links in the food chain, for a broken link bespeaks of a broken chain. That protection can only be achieved collectively.

The next issues to be examined are in the realm of the non-living. Land is a finite resource. The use of land in ways compatible with the above-noted interests is of paramount importance. If, as we have shown, ultimate authority rests with the people, how is it that no limits are placed upon the duration of private ownership of land? Of course, individual owners are mortal and pass possession on, but corporations are not, and nor are the rights of passage. How can the productivity of land possibly be optimized when there is rampant, uncoordinated exploitation for short-term benefits? Those of us who are fortunate to have possession of land cannot be said to own it in the absolute sense, for through endurance land has a longer and therefore stronger claim upon us than we have upon it. We are temporary custodians, many of whom do not warrant the title caretaker for want of caring. The conflicts of interest between taking care of land and taking a profit are complex and run very deep. To overcome these objections a separation is required between ownership in fee simple, and use. Long-term leases are such a mechanism, land titles being held by the people and land use being primarily by the private sector. In this way abuse may be phased out, speculative capitalization of land due to inflation denied, and overall land use planning facilitated for the benefit of all. The same logic, when applied to other natural resources, yields similar returns, all in the best long-term interests of human society.

Now and Then

A capitalist view of this proposition would likely be that it is a subversive communist plot to take over the world, while a communist would view it as the natural consequence of unrestrained, opportunist, capitalist marketing. What is of greater interest is that while the two systems are diametrically opposed in their means, they agree in principle upon ends: one world under one political system. We may not be ready for then 'now', but we assuredly should not be found wanting now 'then'.

The Way Things Are

Contemporary industrialized society overcomes the isolation of the individual by promoting herd behaviour. In totalitarian societies this is achieved through coercion to effect uniformity of action if not thought. Democratic societies, though more subtle, employ propaganda to induce conformity, the exercise of 'free will' being touted as proof to the contrary, as freedom of choice, albeit conditional. Consensus opinion, however, nurtured and groomed as it is by television, advertising and pollsters, confirms a large degree of homogeneity in behaviour, essential to the smooth running of the institutions of state. Acquiescence to the dictates of the state's peculiar brand of ideology is achieved through the induction of fear, fear of being excluded.

The mechanisms of state, be they capitalist or communist, depend upon the reliability of the citizenry. To this end the state cultivates and rewards those who conform to its espoused goals. The capitalist state pays well on the premise that earnings will be ploughed back into the economy, while the communist state pays poorly on the understanding that it expects little from its members. The effect is much the same insofar as the primary purpose of the state is perpetuated. States seek to continue to be what they are and do what they do with a minimum of disaffection. By both standards the ideal state would be operated by placid, compliant, contented automatons. Modern man has been carefully cultivated to consume with a smile. His character has been conditioned to rely upon his primary senses as the arbiters of satisfaction. Invoking the words of Frank Lloyd Wright, modern man's principle interests are 'three squares, fornication and a good snore'. Man is in danger of undergoing an atavistic reversion to the primate state of his forebears. Granted his ranch-style suburban home has replaced the cave, his three-piece suit and button-down shirt up-stages his loincloth and his Mercedes murks his mule. But he has succumbed to temptation and the cultural code, hook, line and sinker, and is no longer his own man with his own will, defiance and pride. Nor is he a happy man.

In consequence man as worker has become divorced from man as soul-bearer. He is valued by the system of which he finds himself a part, as any other part. As an individual he has become disposable, which is distinctly different from being substituted at the end of his tenure. He has become a consuming and consumable commodity, not unlike a fish in the ocean, except that he has allowed himself of his free will, or failure to exercise it, to cannibalize his conscience.

The ideals of humanism as the standard of communism in the east and of capitalism in the west have been betrayed.

Facilitating Social Change

The remission of fear is the key indicator of success of any social programme. Thus, as each step of progress is recorded a corresponding measure of confidence is infused and a reciprocal measure of self-protection may be withdrawn. If all are defended, who needs guns? If all are provided healthcare services, who needs private healthcare insurance? If all are nourished, who needs to beg? If all are sheltered, who is homeless? If all can read and write, all can heed the right! Where international emergencies threaten our collective security we opt for international cooperation without hesitation. All is within our collective capabilities.

The evaluation of all things in terms of commodities, including human life and aspirations, has had the effect of automating mankind back into a state of less-than-human being. Such a conditioned state of body and mind lends itself to manipulation to the advantage of the few who are the best players of that game, for their material and mental satisfaction. Under such conditions, the two fundamental social aspirations, the wellbeing of people in the singular and in the plural, have been completely negated. What can be done to turn such developments around so that people are liberated from such a fate?

Any westerner who has visited the east is familiar with the difficulty of eliciting an answer to a question of fact. This is not due to unwillingness on the part of those who are born into eastern cultures, on the contrary they are extraordinarily gracious by western standards. It is because they don't understand why one would want to know, quite unrelated to the question itself. The root of this apparent bewilderment emerges from distinctions between the two cultures. In the west social change has evolved incrementally over the last 2.5 millennia, the east has been bludgeoned by a twentieth-century social revolution so profound that, within living memory, societies based on abject serfdom are being transformed into dynamic competitive industrial states with seemingly limitless potential for economic growth. The spectacular transformation in mentality from 'who am I to question why, mine is just to do and die', to 'I can do anything better than you can, I can do anything better than you' confounds Western comprehension.

In a democracy the answer begins, literally, with a renaissance or rebirth of the human spirit. What is essential before anything positive can be done to change the trends described is for there to be a phenomenal awakening of public consciousness of the present condition and the consequences of the continuance of current societal trends. Only then can the public, as individuals and collectively, have a chance to offset selfish hedonistic cultures with something of a higher order, that is, something that aspires to a greater, a more general, collective good than the pursuit of individual pleasure can achieve.

There is a social outlook that characterizes people who are not well integrated into society that is represented by the following: the first person (me) is always good, the second person (you) is OK and third persons (he or she), particularly in the plural (they), are generally suspect, especially in their absence.

The synthesis of the interests of the individual with those of society is incubated and nurtured in the family, later to be supplemented by fraternal groups until the full extent of society is embraced. If this does not occur, neither will integration.

In our world few can impact the status quo to advance social improvement. In the ideal world all of us could do so. So all we need to do is to move incrementally in the direction of the ideal world, not to reach the ideal, for that is inconceivable, but to effect social improvement as the primary and continuing social purpose. The path to that end depends upon our collective will and efforts applied to redirecting the social compact from emphasis on economic capitalism to emphasis on humane human capitalism. Only in humanitarian conduct lie opportunities for all to participate in advancing social improvement.

If democracy cannot achieve these objectives, the alternative, at no small risk to society, is to pass the responsibility on to a minority, an aristocracy (or rule of the best) that can demonstrate that it has a better way. The best outcome of such an alternative would be an enlightened meritocracy under a modern equivalent of Plato's philosopher-king, the worst a self-serving dictatorship. A betting man would hedge his bets and opt back to democracy. Philosopher-kings have been known to resuscitate failing city-states, and in contemporary commercial life large corporate enterprises, while dictators have brought the world to its knees more than once in the twentieth century. Perhaps a philosopher-king under the umbrella of a constitution and democratic veto could perform the requisite miracles. Bees function with a degree of leverage to their social advantage unimagined in human society.

56   Sexuality: Repercussions of Living

Nature has distinctive ways of perpetuating life: division, which is a more urbane name for subtraction, a process known to biology as asexual reproduction, multiplication, which is addition seen through rose-coloured spectacles and biologically dubbed sexual reproduction; plus a few surprising predilections that we shall discover in due course.

Homogeneity

In asexual reproduction a simple single-cell organism divides to become two ostensibly identical twins. An amoeba is of this class. By literally acquiring the same gene pool as the single parent, the offspring enjoys all of its traits. The quality of the inheritance is not diminished or enhanced from that of the parent, and their offspring will enjoy the same prospects. They have no intrinsic abilities to mutate away from being what they are. To evolve they depend solely upon extrinsic environmental impacts.

Hermaphroditism

There are, however, many species of plants and animals that practice a kind of do-it-(to)-yourself fertilization act by virtue of combining male and female characteristics in the same organism, some practising self-impregnation.

In the botanical world hermaphrodites include all flowering plants that carry the stamen and pistil in the same flower, of which the coltsfoot and the carline thistle are examples, and those that extend flagella (long slender shoots or roots) as a means by which to propagate, of which group the strawberry is a member.

Zoologists classify many invertebrates as hermaphroditic. These embrace marine molluscs, including shellfish and limpets, and certain asymmetrical flat fish that produce eggs and sperm in the same individual. On land hermaphroditic molluscs include slugs and snails, while earthworms and certain parasitic worms, e.g. tapeworms and flukes, occasionally engaging in autocopulation, all belong to this general class.

Hermaphrodites suffer the same deficiency as asexual organisms in that they bear no inherent capacity to mutate and thereby reconcile themselves to evolving environments. Such a limitation weighs probability heavily against the regeneration of such species in perpetuity.

Heterogeneity

Sexual reproduction is the means by which the more complex living organisms maintain their kind. As a consequence of commingling the genetic reservoirs of two parents, the prodigy veers away from the path of identical replication. For example, the estimated 40,000 genes of one human parent can mix in combinations with those of the other to produce great numerical variety. The recipients of certain combinations will inherit faculties that will better enable them to adapt to changes in their environments than could their parents. Those blessed with such enhanced performance will have enhanced survival prospects. Thus, sexuality is a mathematical formulation by which the probability of survival of a species is greatly improved.

Sexuality and Relations

'If I was the only girl in the world, and you were the only boy': do you remember that song? It is about a sexual relationship, not because the girl and boy are sharing or contemplating intimacy, but because they 'exist' as defined earlier and therefore have a relationship, and because their fundamental distinctions in type, as between electron and nucleus, generate attraction of a sexual nature. Broadly stated, this physical reality places every man in the situation of having a simultaneous sexual relationship with every woman, and vice–versa. The potential for gratification between any one man and any one woman in a crowded city is very small, far less between people on opposite sides of the world; but their mutual relations exist, the nature of their relations is physical, and the special nature of their relations is difference in sexual polarity. While these sexual relations may be seen to exist as established by logic, the sexual connection that we make between parties in common understanding is based upon proximate contact and will, both of which are functions of probability. What we call sexual relations are physical contacts of a sexual nature, which is a distinctly narrower construction put upon a relationship for the purpose of inferring intimacy.

A Paradigm for Procreation

Human sexual attraction is like gravity and operates on similar principles in that the strength or value of the forces of attraction appear to decrease proportionate to the square of the distance between bodies, other things, e.g. exposure to view, imagination and prior knowledge, being constant or disregarded. Naturally, those who are by nature more attractive are more successful in attracting. The less attractive must work harder on the 'other things' in inverse proportion to their natural attractiveness. To this end the human body parts are used for advertising, subtly or otherwise, to arouse sexual interest and to procure sexual opportunities.

Sensuality

Sexuality is unconscious in plants and subconscious in animals, with one exception. While sexuality serves mankind in the same way that it serves other animals, advancing interests in survival, the awakening of the human mind to conscious thought generates confusion. Mankind's ability to anticipate and plan sexual activity extends it from a brief encounter to becoming a preoccupation. In anticipation we derive pleasure that we would like to prolong, even vicariously when we have no partner. Indeed, for some people such contemplation consumes many of their wakeful hours.

Taking one's pleasure in this way transforms purpose, subordinating reproduction to a secondary effect rather than a primary cause. While procreation is not the immediate incentive motivating animals to mate, they do so without premeditation. Man is fully aware of the consequences of his actions yet chooses to act so solely to gain sensual gratification. In this sense his sexual behaviour is an aberration from that naturally indulged by other animals. Such conduct has become the norm by which human sexuality is practised, thereby no longer qualifying as an aberration.

Aberrant Sexual Behaviour

Confining sexual interest to sensory pleasure renders the step in logic to single-sex relations a practical alternative, safe from unwanted pregnancies and long-term commitments, and offering a path to slowing rates of human population increase. If this variant were to gain popular acceptance, it too would no longer be held to be aberrant.

There are, however, certain proprieties that most people prefer to uphold. Self-interest informs man that exposure to extreme risk is inherently dangerous, thus prompting restraint and limiting consideration of such acts to contemplation in the abstract. Homosexuality, with its proclivity for disease transmission, is one such experience, understanding that attraction to such a predilection, like all forms of indecency, can be addictively compelling. Those offended by homosexuality may take comfort that it cannot lead to reproduction in kind.

The logic for what might be called 'virtual', distinguished from 'virtuous' sex, if extended, would cross special boundaries, as between species, and embrace mechanical and electronic means of simulating sexual gratification. The only thing that distinguishes the sexual behaviour of man from that of a beast is that beasts have no knowledge of their actions whereas man does not always act upon the knowledge of his be(a)st interests.

The pursuit of sensual pleasure is superficial and self-indulgent, thereby not warranting extensive commitments of time, virtual or otherwise.

Intemperance

The sexual urge is a terrible distraction from the well-tempered life. It imposes itself upon the consciousness, often at the most inconvenient times. It exhausts the body and diverts the mind to trivial pursuits that are more detrimental to the promotion of wholesome living than they are contributory, excepting the psychological effect of consummate sexual activity relieving stress and therefore contributing to social cohesion. Certain classes of primate use the sexual act much as human beings shake hands, as a means by which to reduce tension upon acquaintance and thereby advance social bonding.

While addressing sexuality and psychology we should note the extraordinary attention that the work of Sigmund Freud enjoyed. In hindsight this is attributable as much to the fact that he addressed the one subject that we all hold near and dear, the relation between sexuality and our personal lives, as to any profundities in the contents of his work. As we have recognized, all is in some way related to all else. Any revelations concerning special relations between the automobile or chewing gum and people are simply not going to trigger the same sensitive psychological interest that the subject of sex does. Exposition and interpolation of this human faculty were Freud's peculiar genius.

Genetic Evolution

As simple asexual organisms became more numerous, their differing environments caused divergence in their evolutionary paths. One branch was to lead to a class of members that developed two separate and distinctively different reproductive organs on the same parent body, an effector and a receptor. Regeneration and the survival of the species depended upon compatibility between the two parts. Advantage lay in the maintenance of simplicity of design of separate mechanisms that performed complementary functions within a single organism in a species that was otherwise becoming progressively more complex. In retrospect hermaphrodites may be seen to have inadvertently evolved a means by which to regenerate gene stocks without the necessity for doing so. Means always precede ends!

The next step in bio-logic was for the attributes of each of the two precedent forms, the asexual (dividers) and the hermaphroditic (multipliers), to combine. By dividing and separating the function of the effector mechanism from that of the receptor, and by placing them in separate organisms, simplicity in modular design was again maintained, while an added advantage accrued whereby a loss of capacity in one mechanism did not preclude the other from performing its reproductive function. The survival advantage would inure to the benefit of those organisms that had the greater capacity to diversify their individual paths to reproductive ends.

The path that sexuality has taken seems to run parallel to the general direction that entropy has dictated whereby organisms have gradually become larger, more complex and more varied.

The Intelligence Quotient

The emergence of the human intellect has delivered much that is beneficial to mankind. Indeed, we have developed to a point where we are no longer simply one of many species being swept along by evolutionary events over which we have little control. We appear to be entering an epoch wherein we have become not merely dominant, but the species upon which all other species depend. It is as if God has elected to share responsibility for the future of life on earth with mankind! In any event, nature's continuous experiment extends!

The question of man's qualifications to perform in such a capacity thus arises. An anomaly presents itself. On the one hand a formidable body of knowledge has been accumulated that, by implication, is essential to underpin any and all significant decisions bearing upon this responsibility. This treasury of knowledge is largely attributable to the initiative and intellectual prowess of individuals. In stark contrast, social advancements proceed at an almost indiscernible rate from one generation to the next. The manner in which mankind addresses its decision-making will reveal whether it has the capacity to assume the challenge of world stewardship.

We shall examine the subject of self-interest later to establish that its primary focus is upon survival in the short term and upon needs that have become attenuated to include 'wants'. Strategies for achieving these goals have been finessed with ingenuity in the interest of delivering gratifying rewards faster. The advent of acceleration in the frequency of changes in our environment is a key indicator that suggests that reliance upon conventional evolutionary mechanisms, including conventional decision-making, may not be sufficient to deliver the necessary capacity to adjust, act responsibly and thereby survive.

While intelligence modifies and directs behaviour in response to changing circumstances in ways that involuntary reflex reaction cannot, what, we need ask, directs intelligence? The quick and dirty answer to this question is our system of values, but underlying this answer is the necessity to differentiate between individual and collective values and to defer to the more compelling interest. Only through the enhancement of probability of human survival can we enhance the probability of human survival. Such is our social imperative. We are thus obliged to adopt courses of action that favour survival of all species as those options most likely to enhance our own survival prospects. Any viable model of societal decision-making is thereby bound to hinge upon the adoption of this proposition. We are obliged to conclude that morality not only 'should' but 'must' direct intelligence in order to deliver the greatest prospects of survival of our species.

Of Mules and Men

By natural selection a process of trial and error has rendered many exotic forms untenable. Upon occasion there come into being the strangest of hybrids, organisms of low probability and little prospect of survival that sometimes owe their homogenesis to inadvertent acts of mankind. Trial and error will likely produce some conspicuous failures, but according to nature's plan, diversity offers the greatest prospects for adaptability and survival of species.

If, as may be contended, man is left to exercise his judgement conventionally, are his demonstrated intellectual and social decision-making capacities up to the challenges ahead? Can he further enhance his prospects of survival by taking unconventional steps towards improving the odds in his favour? What does it take to enable mankind to do so?

The weak link in any chain of events necessary to improve our survival prospects appears to be reliance upon conventional decision-making, whether autocratic or democratic. It is as if we must depart from our conventional 'senses' in the sense that we cannot afford to make survival-critical judgements subject to prejudice and historical belief systems if we are to maximize our prospects of survival.

We are drawn inevitably back to the subject of morality. In the absence of a uniform system of morality we must defer to the exercise of reason. The fuel of reason is knowledge, and the virtue of knowledge its demonstrability. Only through an ability to demonstrate the correctness of our actions can we truly account for them. The intellect has revealed such a process.

Genetic Engineering

What is the next step in sexuality that has potential for delivering significant survival advantage? If, as has been demonstrated, the human constitution is a matrix of interactive genes that elect to switch on or off at a subconscious level, can we enhance human performance by consciously intervening in this process? This is a question that science has an opportunity to address but which conventional wisdom cannot. Of course there will be miscarriages, but can they be any worse than those that we wilfully inflict upon our own kind through the instrumentation of war? Are the moral prerogatives of genetic intervention less humanitarian? The prospects of intervention in the systemic evolution of man are at hand. Morality is on trial. Where lies the guilt?

57   Sensation: Awareness of Existence

To the extent that a nail reacts to the force of a hammer, can the nail be said to have sensed the existence of the hammer? Does this make sense? Does the descent of a cannon ball following its ascent acknowledge the existence of gravity, its own mass or the relation between the two? Does a rock confirm the presence of heat by expanding under its influence? The answer to all of the above questions is no because, by definition, sensation requires the stimulation of a receptor mechanism.

Sensation is confirmation of something. Sensory systems are the means by which that first step towards knowledge of existence is made possible. Sensation does not confirm identity *per se*.

Physiological Sense

Having gathered data confirming the existence of something, does the data make sense? No. The sensation has made data. There remains the need to make the data intelligible; i.e. to translate it into what is commonly called 'sense' but is not. The translation is performed by a related organ, the brain. The brain screens the data for matches with prior experience in order that the 'something' may be classified and labelled. Classification establishes its place in perception and a label its place in language. The 'music' of sensation is impacted upon an organ, relayed through sensory nerves and 'played' to an audience of one in Plato's boundless *sensorium*, the sensory area of the brain. Perception is the score. Without such detection it is impossible to perceive causes or account for effects.

Tradition recognizes five sensory systems of detection: vision, audition, olfaction, gustation and conjunction. These organic detectives discriminate between sight, sound, smell, taste and touch. Tradition also recognizes that there are other sensations that are not so readily accounted for. The sensory systems responsible for mediating 'feelings' of heat, pain, appetite and balance, for example, are not so obvious, and those for tracing 'feelings' of emotion, wellbeing, justice and desire are less so. Even more obscure is the process by which intuitive 'feelings' are 'sensed', for which the term 'sixth sense' is reserved.

Thus, we may ask what the senses are really doing. They are monitoring environmental conditions and reporting back to the boss along established lines of communication. Why should we place limitations upon the number of collector types? Well, perhaps we don't. Perhaps we just distinguish between those that clearly detect and relay different forms of messages and add their numbers. The answer is academic. Academia splits hairs. From a pragmatic standpoint, if it is useful to discriminate between black and white in 3 or 23 tones of grey, we do so. So it is with our sensory systems, the justification for making a distinction between the chemical receptors used to taste and smell being admirably demonstrated by the durian fruit that 'tastes like heaven' and 'smells like hell'.

We return to our prior recognition that senses do not elucidate knowledge, they merely detect presence. By the same token any presence that is not detected, and any new knowledge that is made distinguishable as a function of interpretation, is not by such failure or operation a function of sensation. Insofar as non-organic systems may incorporate receptor mechanisms, sensation cannot by definition be confined to life-bearing or, more narrowly, to conscious phenomena.

Psychological Sense

To the extent that we commonly use the term 'sense' in referring to a detection system, to the transmission and interpretive processes, and to the end product of perception, we invite ambiguity. We further confuse processes of sensation with their products by giving them the same name, as 'to smell' (verb) and sense of 'smell' (noun), that may be confused with a particular 'smell' (noun), being that which is sensed. As a result we seldom list the five traditional sensory systems using the same grammatical forms, or we string together nouns formed from suffixed verbs, as in 'seeing, hearing, smelling, tasting and touching'.

All such uses of the same term to convey different meanings are understandable and excusable when we recognize that they all refer to different characteristics of the same neurological system, a system that knows us better than we know it.

In this discussion the term 'sense' is confined to the process of confirming existence, subjective responsiveness being discussed later under the heading Perception.

Common Sense

As we have seen, common usage is not of itself a remedy for overcoming common misunderstandings. But in time common usage gains official recognition and acceptance into the lexicon, transforming misunderstanding into understanding through qualifying literacy, particularly, as we shall further discuss, through the adoption of a common tongue or *lingua franca*. Explanation makes common sense out of common nonsense, a kind of evolutionary rhetorical justice, but only if one allows the process to broaden the literal meaning of the word 'sense'.

Uncommon Sense

The notion of sense as coherence, the product of intellect, takes us away from our definition of sense as that which detects existence. In theory almost all things are sensible, i.e. detectable, the question as to whether they will be sensed depends upon the presence of an appropriate sensing mechanism. Any novel sensory system is likely to be man-made and target-specific, as is the case where an X-ray machine is manufactured for detecting the presence or absence of cancer. That makes uncommon sense!

Extraordinary Sense

Just as there is the sensible world with which we have direct contact and that delivers all from the most painful to the most illuminating experience, there is also an ideal world, to which love belongs, with which we make comparisons but which we can only discern extrasensorially. Which is more 'real', the sensible or the intangible, is a function of the subject, the object and the circumstance.

Into this latter class of sensation we must place social consciousness. It is as if we are slowly and painfully emerging from an interminable stupor whereby, at high noon, when crises of unprecedented gravity loom to threaten our survival, we come to recognize a new sense, that of personal and collective responsibility. If we can convey and inculcate such a powerful idea into the fabric of social consciousness, we can surely claim to have made extraordinary sense!

58   Touch: Tactile Sensation on Contact

Touch: Superficial Stimulation

While all our sensory systems essentially monitor changes in our environs, our sense of touch may be regarded as our most comprehensive proximity warning system since it comprises a network of sensors that covers the full extent of our boundary membranes. It stands guard at all times to inform us of any contact or impending intrusion by contiguous energy and matter.

Because of the comprehensive embrace of the tactile sensory system, 'touched' is the word that we reserve and use as a synonym to denote that we have been significantly influenced by external events, as by kindness or violence, regardless of the sensory system that so informs us.

Because sensation has the potential for being interpreted as bearing beneficial or negative data, or of informing us of events of little or no consequence, and because potentially negative impacts are of our most immediate concern, we utilize our sense of touch to quickly discriminate between these primary interests.

The information that any sensory system delivers may vary markedly. In the case of tactility; texture, hardness, sharpness and temperature are all expressed in relative terms within broad ranges.

Negative impacts have the effect of stimulating our natural defence systems while positive impacts serve to tranquilize the metabolism. Knowledge of when to resist and when to accept proximate relations is the critical information that we seek.

Taste: The Flavour of Adjacency

In admitting matter within the bounds of our monitoring system for the purpose of replenishing expended energy and matter, we supplement the function of our tactile sensory network with a palatal system that discriminates on the basis of taste.

The function of the gustatory system is to discriminate in the intake of food on the basis of prior acceptability. These tests, while chemical in nature, serve as preliminary determinants as to whether a particular food being offered conforms to the standards of acceptability established by precedence. Unfamiliar food, or that with a benign flavour, tends to fail or pass this test on the basis of superficial likeness to prior experience, exhibiting sweetness or sourness, acidity or alkalinity, excessive heat or cold, and kinship flavours with those of other foods.

Thus, the taste test, like its aesthetic cousin the 'style' test, is an attempt to distinguish the novel from the familiar on the basis of superficial prejudice. Such acceptance or rejection tends to protect the organism from unwarranted intrusions into its more vulnerable inner organs that sequence their own means of rejecting incompatible presentations.

Smell: The Fragrance of Proximity

Insofar as food and fresh air become commingled, the sensation of smell serves to assess the acceptability of airborne particulates on the basis of familiarity. Like gustation, the olfactory system tests are of a chemical nature whereby classes of airborne particles are prejudged and accepted or rejected prior to exposure to the delicate tissues of the respiratory system.

Smell and taste are complementary whereby food offerings may be assessed by both systems simultaneously and correlated before authorizing further acceptance. Insofar as knowledge is the prime determinant of a decision to accept or reject, the connoisseur has a better prospect of making a sound judgement.

59   Hearing: The Faculty of Sound Discrimination

In the preceding discussion of what we have called contact sensors we were able to address the subjects and objects of sensation simultaneously by focusing our attention upon the points where they converge.

Over millions of years animal species have also evolved acute sensibilities to physical events that occur at separate sites, often at great distances from the individuals that sense them. While novel evolutionary developments occur due to improbable circumstances, some characteristics of such events get retained and elaborated upon. In some cases these changes have bestowed added prospects of survival upon an organism, including the capacity to convey such advantages on to succeeding generations. Such is the case with respect to sensitivity to sound, but before we discuss any sensory system we need to identify the nature of that which it has evolved to discern.

Compression Transmission

The release of pressure and consequent dissipation of force is characterized by a succession of waves that transmit energy through a medium with variable intensity. When a force is conducted through a dense medium we assume the velocity of transmission to be instantaneous, as when a hammer strikes a nail. When forces are conducted through a medium less dense, say water, the impact is dampened, as if transmission necessitates the compression of interstitial spaces between molecules that are less densely packed than those of a solid, thereby requiring time to take up the slack. When forces are transmitted through the atmosphere the delay is even more pronounced but for similar reasons. These distinctions enable us to identify natural transmission characteristics for different media.

The compression of a medium implies movement; however, while the wave action is detectable, the medium is not being moved in the direction or at the velocity of the waves but rather vibrates proportionate to the amount of mechanical energy being transmitted. The greater the intensity of energy, the higher the frequency or wave count (and the shorter the wavelength) measured over a given distance. The distance travelled by a wave in a unit of time is described as its velocity and is constant within any medium of uniform density. However, the power of compression waves decays with distance due to the dissipation of a finite amount of energy over an increasing area as waves radiate further from their source. This progressive disbursement of energy translates to a reduction in frequency and an increase in wavelength until, with sufficient resistance, elapsed time and travel distance, the wave action abates and the medium returns to its former state. Radiant wave action is the principle means by which nature redresses the uneven distribution of energy.

In 1842 an Austrian physicist confirmed that the frequency of sound, light and radio wave lengths diminishes proportionate to the rate at which a listener or observer moves away from the source, the converse also being true. Called the Doppler Effect after its discoverer Christian Johann Doppler, the audible form of this phenomenon is an everyday experience sensed in changes in the pitch of sound as the distance between the sound source and the listener changes.

The same attributes of wave action, velocity, intensity, pitch, frequency, wavelength, period, volume, reflection, refraction and polarization attend the transmission of radio, light, heat and sound energy operating within different wave frequency bands. In 1887 the correct correlation between these bands was deduced by Heinrich Hertz whose name is venerated in the unit of measurement used for frequency: one hertz equals one cycle per second, one megahertz one million cycles per second.

This raises the question as to whether all these signals may not be reporting on the same energy disbursement activity but sometimes misunderstood as being separate phenomena by virtue of their being detected by different sensory systems that have evolved independently to detect events occurring within different frequency bands.

The Sound of Pressure

The faculty of hearing has evolved from a rudimentary ability to detect pressure changes within an environmental medium. For respirers that medium is the atmosphere.

The mechanism comprises a means by which to receive ambient pressure waves, a means of transposing them into vibratory impulses and a fibril communication line to the brain for signal interpretation and perception. The sensitivity of the human auditory system is so acute that the approach of a mosquito in flight can be detected by an increase in the pitch or audible wave frequency of its wing beats per second.

The Interpretation of Sound

The principle value of the auditory system, like the other sensory systems, is revealed through its service as an early warning system in alarming the owner of impending danger. For many simple animal species the detection of danger or of prospective mates signals the diversity of sounds that hearing translates to useful information.

While sound originates at focal points and is disbursed through radiant wave action, hearing reverses that process by reducing ambient vibrations to a system of phonetic inflections. Sound judgement is the function of the command and control centre of the neurological system: the brain. Upon receipt the brain compares new sounds with its recorded inventory of prior sounds for the purpose of establishing likeness or novelty from whence it can determine an appropriate response, if any. The ear, thus, functions as an instrument of auditory discrimination, distinguishing alien sound from noise much as we differentiate flowers from weeds – on grounds of their being acceptable! Each sound and combination of sounds is accorded a 'meaning'. The form that meaningfulness takes we shall explore in a later hearing.

60 Light: The Visible Range within the Electromagnetic Spectrum

Precedent to a discussion of sight, we need to know a little about the medium that sight senses.

Light: A Matter of Energy

The subject of light conjures up a profusion of related ideas: energy, matter, radiation, colour, sight, perception and all the wonders that flow into our experience through them. When one attempts to focus on just what light is, a chimerical veil obscures the truth in ways that allow us to gain glimpses without affording us the whole picture. Light is the model format for examining the radiant distribution of energy, because light facilitates sight.

Light as Energy

The peculiar characteristic of radiant energy is that it travels in wavy straight lines. The quality of waviness may at first blush appear to be contradictory to straightness but the waves refer to the mode of its beams that oscillate at high frequencies, the straightness to its general direction of travel. Electromagnetism is depicted as a broad spectral range of comparative waviness. Wavelength is the distance from the crest of one wave to the crest of the next, while frequency describes the number of waves that pass a fixed point in a measure of time. We make distinctions between bands of differing wave frequency on the basis of the means by which we sense them and from our conclusions as to what they do. Thus, at the low-frequency, low-power end of the electromagnetic spectrum and progressively increasing in frequency we have radio, TV, radar, microwave and infrared waves. At the high-frequency, high-power end of the spectrum and progressively decreasing in frequency we have gamma and X-rays and ultraviolet waves. Between the invisible infrared and ultraviolet wave bands the narrow and familiar 'rainbow-coloured' band of visible light ranges from red to violet. It is visible because that is the range within the electromagnetic spectrum that the eye has evolved to detect.

Light as Matter

As a matter of principle all things comprise 'something'. So what is the something of which light is substantially composed? The answer is revealed as a couple of oddities: photons and mesons.

The photon is a discrete package of waste energy that assumes its own identity as it is cast off a decaying atom that loses power due to a change in orbit of an electron from a higher to a lower altitude as the electron moves around an atom's nucleus. Since the photon is a fragment of another particle, it is itself a smaller particle. This casting-off process is the source of light manifest in the sun and stars. The density of so many photons, created simultaneously, is such that they can only move in straight paths away from their parent atoms, deviation being subject to Pauli's exclusion principle that precludes two particles from occupying the same space at the same time. The amplitude or depth of the wave is an expression of how much room each photon has to 'twinkle' its way out of the pack, and how much energy propels it. As a photon proceeds further from its place of 'birth', it garners more room to jostle in proportion to Newton's rule of the square of the distance. Thus, a receding source of light appears to get smaller because less light is reaching a given point, say the eye, proportionate to the distance of that point from the light source. A faintly visible star has the power equivalence of approximately one-100,000th of one watt. The fact that we can detect such a small quantity of light at all reflects upon the discerning sensitivity of the eye.

The meson particle (first recognized as a kind of cosmic ray surfer) bears mass but is unstable since it may carry either a positive, negative or neutral charge. This 'thing', presenting itself in many guises, shuttles between other particles alternating their polarities, as between protons and neutrons in a nucleus, or electrons and protons forming cosmic rays (another form of radiant energy). Since a photon is a spin-off from an electron, the question arises whether any mesons were associated with the electron at the time of its spin-off, and whether any mesons 'escaped' to accompany the newly formed photon.

So now we have particles surfing waves, the cause of much controversy as to whether light comprised particles or waves when it now appears to comprise both a process and a product. To focus solely upon waves and to ignore particles, or otherwise, is to view the phenomenon of light from limited aspects, as different 'dimensions' or ways of looking at the same thing.

Travelling Light

What is this quality of radiant emanation that bespeaks of illumination, and why does it radiate? Radiation is, simply stated, an expression of the flow of energy from areas of high density to areas of lesser density, all else being equal. In this respect it may be viewed as acting independently from gravity. What conditions would support such behaviour? The answer: lack of mass and therefore lack of attraction towards mass.

We recall that Einstein had concluded that as a body loses energy in the form of light, the mass of the body would reduce proportionate to the amount of energy lost. Thus, the mass will equal the energy divided by the velocity of light squared, or $M = E/c^2$. Many will recognize this as his now-famous formula $E = Mc^2$ viewed from another direction. If all is light, nothing is mass.

If the energy/matter is in the form of little modules rather than a single entity, then each is free to seek freedom in the most direct way, i.e. in a straight line from where it is most concentrated to where it meets with the least resistance. As for illumination, that came much later, and is a function of sight and perception rather than light, and so comes later in our discourse.

The Speed of Light

The traditional common-sense view of the speed of light has been that if light passing between two points is measured separately by two observers using reliable means, the results can be expected to correspond. The idea that like-conditions lead to like-results has long been held to be a fundamental precept underlying scientific enquiry into the nature of truth. In free space the velocity of light is not contingent upon wavelength and has been found, through multiple experiments over more than a century, to be universal and constant. The speed of light appears to be the same to every observer.

Pegged at a speed of 299,792.5 km/sec, light waves radiate in one-47th of a second from a point source to form a sphere with a radius equal to that of the earth. The velocity of light varies as it passes through different media due to differing degrees of resistance encountered as native properties of each medium.

Isaac Newton dispelled the idea of an absolute standard of rest by showing that it was impossible to make a distinction between '*A*' moving around '*B*' and '*B*' moving around '*A*'. This revelation necessitated that in stating any speed one would have to stipulate what the speed was measured relative to.

In order to establish a benchmark against which movement could be measured, the idea of an omnipresent 'aether' that occupied all space (including that occupied by matter) was established. Functioning as a theoretical universal coordinate system, the aether would provide a medium for the physical transmission of forces from one body to a second located at a distance from the first. The constant speed of light could then be confirmed by measurement against the aether and observers moving towards or away from the source of light would experience decelerations or accelerations in the speed of light. However, Doppler's discovery brought into question the relation of an observer to a source since this had a critical effect upon the quality of the reception.

In 1887, in an experiment by Albert Michelson and Edward Morley, the conjecture that movement of an observer towards or away from the source of light would effectively change the speed of light was found to be false. Only wave frequencies changed with changes in distance between a source and an observer, as we have observed in our discussion of sound. In defining what speed was to be measured relative to, the observer's time, time became inextricably linked to the observer.

The Doppler Effect applied by analogy to the light of distant stars revealed wavelengths longer than normal. The light tended to redden in appearance, a phenomenon known as the red shift, from which, in 1917, Willem de Sitter construed that distant stars were receding from the earth, and upon which the Big Bang theory of universal expansion was founded.

With general agreement on the speed of light and general disagreement on the distance from the source to an observer due to movement, the time of transmission necessarily remained uncertain. Newton had disposed of absolute place, and Einstein of absolute time.

In the face of this, how can we contend and confirm the reality of absolute speed?

The Standard of Standards

There is a problem in adopting any standard for any measurement and it is of this nature: if the basis of a standard changes relative to what it was, then all measurements relying upon that earlier standard must necessarily change also. Conversion from the statute mile to the metric kilometre may come to the mind of the reader with recognition that while the numbers change, distances do not. The conversion is manageable. But there is something fundamentally more profound about entertaining the idea of a change in the measuring stick of measuring sticks.

If the speed of light, currently assumed to be constant, changed, we would not know about it because we are measuring all other speeds relative to it. Why would it change? For the same reason that everything changes, or rather because we have concluded that everything changes and cannot therefore justify why the speed of light should not. If we assure ourselves through experiment that the speed of light does not vary, how can we reconcile our belief that light is a particle, and that it is massless, when other particles have mass and therefore cannot attain the absolute maximum speed?

A constant speed of light is illogical. It is constant because of the peculiar form of logic that we apply in measuring it, the results of which support our experience and, therefore, our conclusion that the speed of light is constant. Could it be that the reason that the speed of light is agreed upon as being constant is because it is constantly being measured relative to the speed of light that is assumed to be constant? How constant is the length of a one-foot rule if always measured using a twelve-inch rule? How does change in temperature affect change in velocity? Is there a missing factor in our equations? After all, temperature seems to be an essential parameter in determining the velocity of everything else!

We must conclude that we have a working hypothesis that the speed of light does not measurably change, that the hypothesis is useful, and that we should therefore continue to use it as long as it delivers advantages. That is the true measure of worth of a theory, about which we shall entertain further discussion.

61   Colour: The Subdivision of Light

The wavelengths of visible light range from 400–700 millimicrons where 1 millimicron equals one-millionth of a millimetre. That variation is discernible as variety in colour.

Without sight colour is inconceivable. The perception of colour is therefore a process of subjective discrimination between different qualities of light made possible through an ability to discern distinctions in wavelength. Such a means of discrimination is fundamentally different from an ability to distinguish between shape and relative size, also sensed by the eye.

Colour, when viewed as a new dimension of experience, transcends the narrower geometric definition of dimension and in so doing transposes objects from mathematical 'quantities' to become 'qualities' by virtue of endowing subjects with a new format for making comparisons.

It is important to note that the essence of colour is not quality *per se* since it is still prescribed by numbers, in this case quantities or pressure of a particular form of energy bearing upon a particular place in space at a particular time; the individual attaches values to colours.

Correspondence in values between one individual and another depends upon one pair of eyes and one brain sensing and interpreting light precisely in the same way as another. A defective neurological system may result in an inability to translate. Such a condition accounts for partial or total colour-blindness. So colour in the objective sense is unreal.

We are all familiar with the refraction of light exhibited in oil slicks and soap bubbles. Isaac Newton was not the first to recognize that these colours issued from 'white' light, but he was the first to reassemble them and thereby reconstruct 'white' light using a pair of prisms. From this experiment he correctly concluded that all the necessary colours to produce 'white' light were present in the spectrum.

Colour is in the Mind of the Beholder

A prism does not separate the constituents of light by colour, but rather by wavelength. So too do the pigments in paint, flowers and any of a seeming-infinity of so-called coloured objects. Leaves are not green, they merely absorb all frequencies of light except those predominantly in the range of wavelengths between 500–550 millimicrons, which they reflect and we detect and accord the colour green.

When the full range of white (meaning multi-coloured) light passes freely through a medium we are unable to see it. This accounts for our inability to see dry atmosphere, clear water, diamonds and sheet glass. All we see of them are the reflective surfaces at their boundaries.

Just as the eye has a high degree of sensitivity to light, it has a high degree of ability to discriminate between light wavelengths. The 'average' human eye will detect with 'average' acuity the 'average' hue of a colour generated by a wavelength of 605 millimicrons, and owners of the 'average' eyes will likely agree that the colour in question is cadmium red. Those with exceptional powers of discrimination may disagree. We are all automaton 'painters' of the visual images that we 'imagine' we see, imagination being the private world of mental image-makers. Colour pigmented materials owe their distinctive qualities to their inherent abilities to absorb and reflect light of differing wavelengths and to mix, even though neither light nor the materials are coloured *per se*. The resolution of two colours may be found by adding the light wavelengths of each and dividing by two (assuming that their intensities are equal). Thus, cobalt blue (with a 474.6 millimicron wavelength) and zinc yellow (575.8 millimicrons) resolve to become spring green (525.2 millimicrons). With this understanding we can now recognize that by selecting three critical colours of contrasting hues we can mix any intermediate colour by proportionate combination. The three primary colours are not so-called because they enjoy God-given magical powers. They are 'chosen' because they are the frequencies that enable any others within the colour spectrum to be so formulated. A fair analogy would be the stable composition of a triangle whereby, if the lengths and angle of incidence of two sides are known, the 'true colour' (equivalent to the length and orientation of the third side) is inevitably revealed. Since the colours of the spectrum can be represented geometrically as a colour wheel this analogy is particularly poignant; two colours close on the 'wheel' being related through an acute angle, while two colours further separated on the wheel enjoy an obtuse angle relation. The analogy may be further extended by weighting each colour by 'chroma' (strength or weakness) and 'value' (lightness or darkness) to determine a precise specification of the new product of their combination (vis-à-vis its 'length' and 'inclination' equivalence).

The 'quality' of colours belongs to a range of values that we individually place on each, with differing emphasis.

Perceptions of Colour

We will not dwell upon the physics, physiology or psychology of colour here except to note that subjectivity is central to the precept that colour exists at all, and then to the nature of colour experience.

Deferring examination of perception to a later section, we need to briefly acknowledge that the subjective interpretation of colour governs what colour does 'for' people. It differentiates between things and people. The perception of colour, i.e. the mix of wavelengths, is a translatory function of the brain receptors, not of light *per se*, the effect of which is to add a 'dimension' that enables people (and animals) to make distinctions that they could not otherwise make. Thus, colour empowers us to tune in to the environment with greater fidelity than we could otherwise experience without colour, thereby gaining greater insight into the true nature of something. As wavelengths increase, frequencies diminish.

The hallucinatory narcotic mescaline further amplifies the intensity and variety of colours.

What, we may then ask, does colour do 'to' people psychologically? Are greens and blues tranquilizing? Does red provoke ire in bulls and if so in people? What scientific evidence supports such views? Is there, in subjectivity, a built-in bias independent of objective considerations? In order to begin to answer these questions we need to recognize that the eye is much more sensitive to light near the middle of the visual spectrum than near its extremities. At a wavelength of 550 millimicrons the green of new grass and young leaves is midway between the extremes of 400 and 700 millimicrons. The 550 millimicron calibration is analogous to the resting-place of a pendulum in the vertical position, which helps to explain why the colour green is so serene and welcome in the spring after the stresses of winter. Conversely greater application is required to discriminate between the hues of red close to 700 millimicrons in wavelength to which additional stress is attributable.

We have not been endowed with greater acuity at the mid-range of the visible spectrum in order to see better; we see better because our eyes have evolved in an environment that is predominantly green, thereby adapting to serve our best interests in hunting and discerning danger.

How people and colour interact we shall discuss under the subject of aesthetics.

62   Sight: The Vicissitudes of Vision

Having explored sensation generally, acknowledged Newton's celebrated diffraction of coloured light, to which he applied the Latin word spectrum, and having examined so-called white light, that hybrid composite of many colours that renders clarity, we now turn our attention to examining sight, the pre-eminent sensory system that evolved to detect and discriminate between all things bright and beautiful.

Sight places all signals received from lighted stimuli into a relational format, as 'pictures' of the environment. The fact that such captive images transpose up for down and left for right illuminates a function of geometry. It takes brains to unscramble this anomaly.

Stills

We 'photograph' momentary pictures. The reason that we do not simply 'read' these pictures 'live' and use the information directly stems from our need to verify what we are looking at. By scanning a facsimile image and comparing it with pictures that we have previously scanned and stashed in memory, we have the opportunity to confirm whether or not we have seen a similar image before. We are reassuring ourselves. We see a cat, we have seen others; we confirm the presence of a cat by our own definition. From a practical standpoint we do this automatically and so quickly that we cannot recognize a distinction between a 'live' picture and its 'virtual' image.

Movies

The reason that we have evolved such an exotically sophisticated system for looking at cats arises from the fact that cats move in relation to their environments, as we do. We have needs to rely upon knowledge of how relations may change in the environment. Eyes are attuned to pay more attention to movement than to static conditions. Viewing 'stills' is a relatively innocuous use of eyes, employed in search of truth or beauty. Movement signals change, change brings opportunity, opportunities bring rewards and punishment. Those that hunt and are hunted are more competent and less vulnerable if they can sense telltale signs of movement. The more sophisticated the sensory system the greater the probabilities of rewards and the smaller the likelihood of punishment. Survival may well depend upon discerning movement while not being discerned when moving.

While we earlier acknowledged that visual stress was encountered in focusing upon colours at the extremities of the visible spectrum, bull's eyes are more sensitive to movement than to colour. Just as long as matadors stand still and capes move, the matador merits less attention.

The Dark Side of Seeing

Much is invisible simply because our eyes are inadequate to render optical images within the full range of the electromagnetic spectrum. In this sense sight within and beyond the infrared and ultraviolet bandwidths is blindness. We cannot see what we cannot see.

Perspective: A Point of View

Sight has given us insight into our immediate surroundings. The shock of such an introduction is considerable. Not only do we have a confusion of shapes, colours and movement to contend with, but an urgent need to sort them out with respect to priorities of meaning. By meaning we here refer to potential for rendering benefit or disadvantage to the individual. The only sight in sight belongs to the first person singular.

Depth of Field

If we examine a photograph of a familiar landscape we delude ourselves that we see a depth of field, that is, that we see an array of images in which the nearer objects obscure or overlay images of objects further away. This is a delusion because all objects in the photograph are equidistant from our eyes. We are looking at a flat two-dimensional 'picture'.

When we put the photograph aside and behold the view that the photograph represents, we gain a general impression of the whole 'picture' with sufficient sharpness in detail to enable us to distinguish relations between the images in the field. The array of stimuli ranges in distance from the eye from the very close to the very distant. We are said to have a deep depth of field, but we still have a 'flat' image on the retina of the eye.

Necessity to Differentiate Distance

Shape, colour and movement do not afford us enough information about our environment to assure security. Our security is potentially at risk when something external to us threatens to occupy our personal space-time. We therefore need knowledge of the relative proximity of such stimuli, from which we may infer a degree of personal detachment and relative safety.

Comparison by Relative Proximity

Our experience offers us a general understanding that as stimuli recede from us they appear to diminish in size. We revisit Newton, recalling his 'square of the distance' rule as it applied to gravity, to discover that the same rule applies to scale. By this means linear dimensions diminish by half with a doubling of distance, and areas of planes diminish by the square of the distance. Thus, a 'No

Parking' sign appears to be half as high, half as wide and (½ x ½ =) one quarter the surface area of a similar sign one half the distance from the viewer. This experience is born of geometry and the theory of proportionality, whereby the ratio of relative positions is retained with an increase in distance, but appearances of size diminish.

Perspective and parallax deny a sensual appreciation of truly objective Euclidean space. Instead, we must rely upon mental images (theories) of correspondence to true-to-scale images that we see only in diminishing proportions. We now must recognize 'experienced space' to be something other than what we have earlier represented space as being: objective Euclidean, Riemannian, curved or any other mechanically consistent construct.

Reliance upon Memory for Acceptance of Scale

We are born with a deficit of knowledge of myriads of different stimuli that confront us daily. We improvise by imperfectly registering the relative sizes of everything that we encounter and banking this data in memory. Upon reacquaintance we compare the new with the graven images to confirm size, shape and colour in the interest of being consistent with our expectations and thereby establishing a degree of acceptability. If we perceive an image novel to our experience, our memory will assure that it will not be novel upon our next acquaintance. In the absence of comparative scales we are at a loss to instantly distinguish a yawning domesticated cat from a roaring lion.

Physiological Limitations

In addressing scale we should not forget that scale is a function of sight, and sight a function of visible light. Constraints that limit the passage of light, as when one stimulus obscures a view of another by eclipsing lines of visible light from reaching the eye, thereby constrain sight and deny access to perspective. Every shadow is a product of an ecliptic relationship, a coincidence of alignment. Had we been endowed with X-ray vision or the ability to decipher synchronized laser light, we would be blessed with the faculty of 'seeing' things in perspective through other things, and could 'ghost' three-dimensional holographic images, but would likely be confused by a surfeit of information.

Perspective as Metaphor

The term 'perspective' lends itself to use in many ways. We use the term to convey a particular vision of the physical environment, to represent a graphic construct of relations, as in a painting, as a prospect of the future, and, lastly, to recognize sagacity or intellectual capacity.

Psychological Limitations

In 1884 a staid English schoolmaster, Cambridge-educated in theology and ordained into the ministry, published his first and only novel. *Flatland* is the story of a world of two dimensions in which fictitious character types are identifiable by their geometric shapes. The hierarchy of this culture, a satire on Victorian English society, comprises the lower classes representing all women as straight lines, soldiers as acute-angled isosceles triangles, workmen similarly but with less acute angles and merchants as equilateral triangles. The upper classes comprise professional men portrayed as squares, gentlemen as pentagons, noblemen as hexagons and at the top of the class progression the high priest, a multi-faceted polygon almost indistinguishable from a circle; all carrying their badges of relative social standing in the number of facets to their form.

Significantly, the main character of the story is named '*A* Square'. Written by a member of the learned professions, the story can perhaps be viewed as autobiographically inspired. The unusual duplication in the author's name, Edwin Abbott Abbott, conjures up images of the geometric concept *A* to the power of two, or *A* squared (A^2), suggesting grounds for a special personal identification with the difficulties that '*A* Square' experiences in the story.

'*A* Square' has a dream that he visits a fictitious 'Lineland' where he encounters a multitude of lines of various lengths, moving to and fro along a single alignment. Naturally, he assumes them to be women, whereupon he attempts to engage the longest in conversation, only to be brusquely informed that he is addressing the King, clearly so by virtue of his length. In 'Lineland' the women are simple points. In further discussion, '*A* Square' attempts unsuccessfully to introduce the King to the concept of two dimensions by passing across his field of vision, but all the King sees is a point that extends to a line and then diminishes to a point again, and vanishes; a kind of sectionalized view of a square as it passes through an infinitely small aperture-like field of vision. The tables are turned on '*A* Square' the next day when he is visited by a point that transmutes itself into progressively enlarging circles, diminishing by the reverse process, and disappearing. It is '*A* Sphere' coming to teach '*A* Square' about the third dimension. '*A* Square' endeavours to acquaint his peers with ideas of other dimensions, drawing upon the analogy that a single-dimensional figure, a straight line (or woman), cannot readily comprehend the experiential world of two-dimensional figures (men). When '*A* Square' attempts to convince his fellow 2Ds of a third dimension he is promptly thrown in jail.

The analogy carries over to illustrate our own difficulties in comprehending a fourth and higher dimensions, whether mathematical, physical or metaphysical. However, each additional sensory system secures for its bearer new insights into the true nature of its environment.

It is noteworthy that the animation of a 'thing', point, line or plane serves to heighten the identification of a reader or observer with the thing under consideration, a valuable means by which to gain consciousness of the deeper nature of that 'thing'.

We shall close this brief encounter with the senses by observing that, upon occasion, any or all of our primary senses can be brought into convergence by a single event. Such may be our compounded sensation of the seashore. Here sounds and sights of repetitive pressure waves are overlaid by the rhythmic contact of rising and falling water levels, made more piquant by the complex mixture of tastes and smells of animal, plant and mineral origins.

Such is the wide range of variable performance of the sensory systems that each is capable of inducing the most tranquil of dispositions and the most petrifying of traumas in our minds.

Progressive development of the sensory organs has culminated in the most complex evolutionary development to date, intellect, and a means by which individual intellects can connect through language.

Index